THE GARDEN OF EDEN

THE
GARDEN OF EDEN

The Botanic Garden and the Re-Creation of Paradise

JOHN PREST

YALE UNIVERSITY PRESS

NEW HAVEN AND LONDON

1. (frontispiece) The entrance gate to the Botanic Garden
at Oxford, dedicated in honour of King Charles I to the
glory of God for the benefit of the university and of the
nation. The first gardener, Jacob Bobart the elder, stands
outside together with his goat and his dog. The stork is
probably an allusion to Bobart's coming from a part of
Northern Europe where storks nested in large numbers,
and the serpent on a staff was a symbol of herbal
knowledge.
'Vertumnus' *An Epistle to Mr Jacob Bobart*, 1713

Designed by John Nicoll
and set in Monophoto Baskerville.
Printed in Great Britain by
Jolly & Barber Limited,
Rugby, Warwickshire.

Published in Great Britain, Europe, Africa, and
Asia (except Japan) by Yale University Press
Ltd., London. Distributed in Australia and
New Zealand by Book & Film Services, Artarmon,
N.S.W., Australia; and in Japan by Harper & Row
Publishers, Tokyo Office.

Library of Congress Cataloging in Publication Data

Prest, John M.
The Garden of Eden.
Bibliography: p.
Includes index.
1. Botanical gardens – Europe – History.
I. Title.

QK73.E85P73	712'.5	81–11365
ISBN 0–300–02726–5		AACR2
ISBN 0–300–04370–8 (pbk.)		

LIST OF CONTENTS

ACKNOWLEDGEMENTS

I am grateful to the Trustees of the Will of the late Major Peter George Evelyn deceased for permission to refer to the 'Elysium Britannicum', to Ken Burras for advice given in the Botanic Garden, and to both Peter Hinchliff and Keith Thomas for encouragement and correction, going far beyond the claims even of long-standing friendship, given at every stage in the production of this book.

For their courtesy in allowing me to reproduce illustrations I wish to thank: for colour plates; plate I, the Trustees of the Victoria and Albert Museum; plate II, the President and Fellows of Corpus Christi College, Oxford; plate III, the Städelsches Kunstinstitut in Frankfurt am Main; plates IV, VI and VII, the Bodleian Library; plate V, the Trustees of the Ashmolean Museum: for black and white illustrations; 5, J. K. Burras, the Superintendent of the Botanic Garden at Oxford; 10, Rosenkilde and Bagger, International Publishers, Copenhagen; 31, the Rijksmuseum, Amsterdam; 34, the Trustees of the Will of the late Major Peter George Evelyn deceased; 67, *Country Life*; 70, the National Gallery in Prague; and 71 the Worcester Art Museum, Worcester, Mass. The copyright in the negatives of all the remainder of the photographs rests in the Bodleian Library. I wish to thank Mr C. G. Cordeaux and the photographers for their ready help, and to say how much I appreciate being allowed to reproduce nos 54 and 55 taken from the herbarium of the philosopher John Locke. The shelfmarks of the remainder are:

1, Gough Oxon. 109 (11); 2, Sherard 392 (2); 3, 8° C. 23 Med.; 6, Sherard 241; 7, Douce P. subt. 42; 8, 19183 d. 26; 9, 4° P. 43 (2) Th.; 11, 12, Douce A. 302; 13, 19183 d. 26; 14, 15, Douce A. 302; 16, 4° P. 43 (2) Th.; 17, Douce, D. D. 178; 18, Sherard 446; 19, 2594 d. 15 (66); 20, Douce R. 577; 21, Mason A. A. 331; 22, 23, Douce R. 577; 24, R. Pal. 9.24; 25, K. 3. 6 Art.; 26, Douce S. subt. 2; 27, Antiq. f.F. 1691/2; 28, 4° Z 143 Jur.; 29, G 24/B1. 64/3; 30, Douce H. 455; 32, 8° C. 23 Med.; 33, Sherard 241; 35, Sherard 621 (2); 36, 37 Sherard 91; 38, Sherard 404; 39, Douce A. 302; 40, Sherard 701; 41, 42, 8° C. 23 Med.; 43, Sherard 241; 44, Vet. A. 4. b. 12; 45, Douce F. 235; 46, fol. St. Am. 61; 47, Sherard 26; 48, Vet. A. 4. e. 588; 49, 50, Sherard 594; 51, Douce QQ 11 (1); 52, Vet. E.3. b. 12; 53, Sherard 594; 56, Antiq. e.E. 1657. 3; 57, 58, Douce QQ 11 (1); 59, 60, 61, Douce Prints a.24; 62, Vet. A. 4. d. 178; 63, Arch. Antiq. A. II 13; 64, 17356 a. 4; 65, Vet. A. 4. d. 178; 66, Vet. A. 4. d. 53; 68, 69, Gough maps 178.

INTRODUCTION

AT first sight, a Botanic Garden strikes the casual visitor as being quite unlike any other garden. There are flowers, but it is not a flower garden. There are vegetables, but it is not a kitchen garden. There is fruit, but it is not an orchard. There are trees, but it is not a park. There are even some 'weeds' cultivated and treasured like other plants. What, then, is it?

The great age of the Botanic Garden followed the discovery of the New World, and six of the Botanic Gardens founded in Europe achieved a kind of pre-eminence. Three of these gardens, those at Padua, Leyden, and Montpellier, were established in the sixteenth century, and three more, Oxford, Oxford's close contemporary the *Jardin du Roi* at Paris, and Uppsala, in the seventeenth century.

The Oxford Garden was founded by the Earl of Danby in 1621, opened in 1632, and completed in 1640. The first gardener was not John Tradescant, the King's Gardener, as had been hoped, but a discharged soldier from Brunswick, Jacob Bobart, who 'first gave life and beauty to this famous place',[1] and was succeeded by his son, Jacob Bobart the younger. The Garden was, and to a large extent still is, surrounded by a massive stone wall, fourteen feet high, and the visitor used to enter by (he now enters beside) a monumental arch built by Neklaus Stone to a design by Inigo Jones (Plate 1). The weathered arch remains, sadly crowded and overshadowed by the buildings of the old Botany School on one side, and the former residence of the nineteenth century Professors of Botany on the other.

In shape the Garden was square: within, the main lines were formally laid out, and the three acres were divided into four quarters. This plan, which had already, for a period of over two thousand years of gardening history, represented the four corners of the earth, now stood, since the discovery of America, much more specifically, for the four continents, Europe, Asia, Africa, and America, into which the land-surface was known to be divided. Each quarter was sub-divided into separate beds or parterres (Plate 2). In Padua and Paris the parterres in the quarters were laid out in intricate geometrical designs (Plate 3). In Leyden they were arranged according to a simpler plan. There, each bed was long, straight, and narrow, and was known, from the Latin, as a *pulvillus* (or small cushion). At Oxford, the original layout, parts of which bore some resemblance to the ornate designs at Padua and Paris (Plate 4), has long since disappeared, and today the individual beds are more reminiscent of the seventeenth-century arrangements at Leyden (Plate 5).

Among the parterres, each *pulvillus* became the home of a particular family of plants, and each *pulvillus*, in its turn, was further sub-divided into smaller units, which were numbered, where the different members of a family found their individual seats (Plate 6). The plants were thus allotted places, in order, like the members of a great household sitting upon benches on either side of a long dining

SPALDO TERZO.

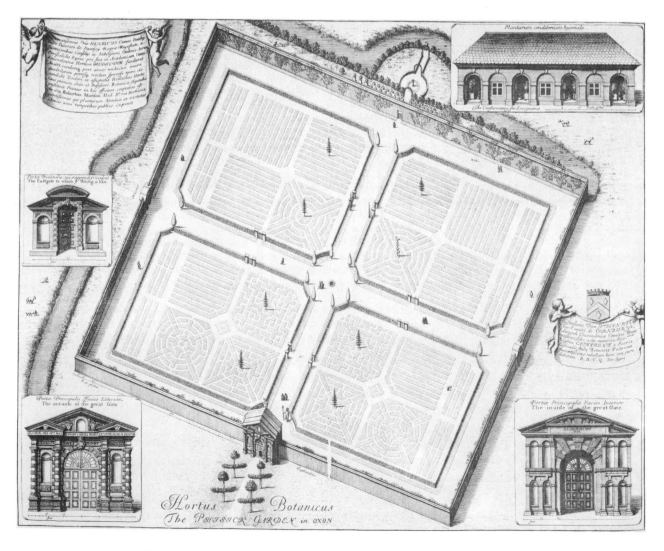

Plantarum conditorium hyemale

The Conservatory for Ever-greens

Portæ Orientalis im respondet Occident.
The East-gate to which ye West is like

Portæ Principalis Facies Exterior.
The out-side of the great Gate

Portæ Principalis Facies Interior
The in-side of the great Gate

Hortus — Botanicus
The PHYSICK GARDEN in OXON

2. (facing page, top left) The four quarters of the formal section of the *Jardin du Roi* at Paris.
G. de la Brosse *Description du Jardin Royal des plantes médecinales*, 1636.

3. (facing page, top right) The third quarter of the Garden at Padua showing both the intricately ordered design and the numbering of the beds.
G. Porro *L'horto de i semplici di Padova*, 1591.

4. (left) Loggan's plan of the Garden at Oxford.
Oxonia illustrata, 1675.

5. (above) The Botanic Garden at Oxford today contains reconstructed *pulvilli* and a fountain in the centre. Mature trees create a paradisal effect. Copyright J. K. Burras.

Colour Plate I. (following pages) Jan Brueghel the elder (1568–1625) presented the Garden of Eden as a place where flowers and fruits appeared on the trees simultaneously, and the animals lived at peace among the glades of a park or landscape. Copyright, the Victoria and Albert Museum.

6. Detail of the Garden at Leyden showing the numbering of the plots in a *pulvillus*.
P. Paaw *Hortus publicus academiae Lugdunum-Batavae*, 1601.

7. (right) Texts from *Psalm 148* and from *Paradise Lost* have been added to this copy of John Parkinson's *Paradisi in sole*, 1629. No animals are shown in this Garden of Eden, but the vegetable kingdom includes the Scythian lamb (see plate 37).

table. And just as there was no place-setting at a dining table that could not be reached from the passageways left between the tables in a dining hall, so, at Leyden in the seventeenth century, there was no plant that could not be seen, touched, smelled, and sketched from the gravel or grass walks between the *pulvilli*.

This feature, more than any other, betrays the academic origins of a Botanic Garden. The Garden was an encyclopaedia. Like an encyclopaedia it was a 'book', laid out in pages, which were 'printed' or 'set' for reference. It had the advantage over a book that the plants were real, and took precedence over a herbarium (or plant museum) of cut, dried, and mounted specimens, because the material was alive. No visit to a modern Botanic Garden can now convey the sense of excitement with which, between three and four hundred years ago, these first encyclopaedias were compiled. Each new plant from the farthest corners of the world was awaited with expectation, and received with enthusiasm, identified, and named.

It might appear, then, that in the Botanic Garden, one stands in the presence of the beginnings of modern science, the collection of data, and the patient, detailed observation of causes and their effects. These are, indeed, the directions in which the Garden ultimately led, but reference to modern science does not describe the motivation with which it began. Contemporaries interpreted the foundation of these encyclopaedic Gardens in a context of the re-creation of the earthly Paradise, or Garden of Eden, with which this story begins (Plate 7). There was at that

6

PARADISI IN SOLE
Paradisus Terrestris.
or
A Garden of all sorts of pleasant flowers which our
English ayre will permitt to be noursed vp:
with
A Kitchen garden of all manner of herbes, rootes, & fruites
for meate or sause vsed with vs,
and
An Orchard of all sorte of fruitbearing Trees
and shrubbes fit for our Land
together
With the right ordering planting & preseruing
of them and their vses & vertues
Collected by John Parkinson
Apothecary of London
1656

Qui veut parangonner l'artifice a Nature,
Et nos parcs à l'Eden, indiscret il mesure.
Le pas de l'elephant par le pas du ciron,
Et de l'Hiesle val par cil du moucheron.

Adam hic uestitus e ad laborandū a
eua sedet 2 lactet filiū siū abel 2 stat
iurta eam quasi puer

Filii Adam · Caym · Abel · Seth ·

ffactum est ōmc trīps qᵒ vir et adam · sᵈᶜᶜᶜ xxx An·

aꝑtus minutie

Colour Plate II. (left) This early fifteenth-century manuscript of the *Speculum humanae salvationis* showing Eve spinning, with the cradle at her side, and Adam with his spade, concentrates upon toil as the consequence of the Fall. Copyright, Corpus Christi College, Oxford.

Colour Plate III. (above) The Oberrheinischer Meister's 'little paradise garden' (*c.* 1410) is full of religious allegory, from the Virgin sitting reading, and the Child playing at music with his Mother, to the two trees, and the life-giving cistern. Animals are excluded, but the garden is open to songsters and angels. Copyright, Städelsches Kunstinstitut, Frankfurt am Main.

time no intellectually credible alternative to the account of the origin of the world given in *Genesis*.[2] Catholics and Protestants alike held that the first home of mankind had lain in a Garden planted by God, where the climate was always mild, and the trees flowered and bore fruit continuously. Throughout the middle ages the Garden was believed, somehow, to have survived the Flood, and in the great age of geographical discoveries in the fifteenth century, navigators and explorers had hopes of finding it. When it turned out that neither East nor West Indies contained the Garden of Eden, men began to think, instead, in terms of bringing the scattered pieces of the creation together into a Botanic Garden, or new Garden of Eden.

For the first time since the Fall, thanks to the discovery of America, a truly encyclopaedic collection of plants could now be made that would offer a complete guide to the many faces of the Creator. Since each family of plants was thought to represent a specific act of creation, that scholar would come to understand God best who found room in a *pulvillus* for every *genus*. This, then, was the Garden of re-creation into which the wise man would retire, and shut the door upon the busy, disfigured world outside. Plants were restful things, free from motion, and, so it was generally imagined, from the perturbations of sex. In this they resembled God himself. Not only that: the leaves of the various plants having been appointed by God for physick (or medicine), a complete collection of plants from all over the world, must, it was supposed, supply a ready remedy for every injury and infection. Thus it was that, in a Botanic Garden, beside the fountain in the middle, a man could enter into communion with what was green and full of sap, recover his innocence, and shed his fear of decay.

Some of the brightest hopes of mankind thus came to lie in the principles of re-creational gardening associated with the Botanic Gardens and with what was written about them. But these ideals could be adopted by other gardeners, both great and small, with notions of their own as to the many ways in which to re-create a 'perpetual Spring'. After the relative hopelessness of the medieval world view, and the division of Christendom at the Reformation, these attempts to re-create the Garden of Eden, backward-looking as they were, came as a sensitive, and immensely optimistic interlude in Western European history, before the march of modern science and of the industrial revolution began.

8. God inducts Adam and Eve, who walk in step, into the Garden of Eden. The Garden does contain animals, and emphasis is laid upon the fountain of life and the four rivers issuing from the enclosing wall.

    Ludolphus of Saxonia *Leven Jhesu Christi*, 1503.

CHAPTER I

EDEN

THE two most striking features of the mythical golden ages invented by so many of the early civilisations of the Mediterranean and the middle east were the belief in a perpetual Spring, and the desire for peace with, and among, the animals. Men, compelled to earn their living in the open air, and through the cycle of the seasons, dreamed of a time when the weather had been temperate, and the earth had brought forth its fruits in abundance and without toil all the year round: Homer described the delights of a perpetual Spring, in the garden of Alcinous, where, in a large orchard of four acres 'Their fruit never fails nor runs short, winter and summer alike, it comes at all seasons of the year, and there is never a time when the [soft] West Wind's breath is not assisting, here the bud, and here the ripening fruit; so that pear after pear, apple after apple, cluster on cluster of grapes, and fig upon fig are always coming to perfection.' Societies working with tame, and living in fear of wild animals, cherished the ideal of a truce between the creatures: Isaiah spoke of the day when 'The wolf also shall dwell with the lamb, and the leopard shall lie down with the kid; and the calf and the young lion and the fatling together, and a little child shall lead them.' And Virgil, too, looked forward to the time when the ox would no longer be frightened of the lion.[1]

When we turn to the biblical story of the creation and of the Garden of Eden, it might not at first sight appear, since there is no specific mention of a perpetual Spring, that the Mosaic account shares the same concerns. But there can be no mistaking the fruitfulness of the trees from which the perpetual Spring has always been deduced, and peaceful relations among the animals can be supposed to follow from the fact that God gave 'to every beast of the earth, and to every fowl of the air, and to everything that creepeth upon the earth . . . green herb for meat.'[2] Man also, with his original diet of seeds and fruit, lived at peace with the animals, and the (somewhat one-sided but equable) relationship between them was established when God brought the animals into the Garden for man to name, and commanded them to obey him.[3]

The Mosaic account, then, is no exception to the general pattern. That does not mean, however, that anyone seeking to restore the Garden of Eden, with a mild climate and a harmonious creation, would find *Genesis* an easy book to interpret. Exact details are few. The Garden had boundaries, and there were, at the centre, two trees named as the tree of life and the tree of knowledge of good and evil (Plate 8). Four rivers issued from it, the Euphrates, the Hiddekel or Tigris, the Phison and the Gihon, and there was gold in one of the adjacent countries.[4] And that is all. The statement that it contained 'every tree that is pleasant to the sight and good for food'[5] was generally taken to mean that every plant found a home there, but could be understood to refer to a selection, only, of the best. A majority of the early Fathers interpreted the Garden as a place in which animals wandered freely

9. (right) Adam and Eve stroll arm in arm among the animals in the Garden.
J. Fletcher *The Historie of the Perfect-Cursed-Blessed Man*, 1628.

10. (facing page, above) An eleventh-century impression of God pointing to the forbidden tree, which separates him from Adam, with one hand, and to the animals over whom Adam has been given dominion, with the other.
The Old English Illustrated Hexateuch edited by C. R. Dodwell and Peter Clemoes. Copyright Rosenkilde and Bagger, Copenhagen.

11. (facing page, below) Adam names the animals where they stand just outside the entrance to the Garden.
G. B. Andreini *L'Adamo, sacra rapresentatione*, 1617.

among the trees and over what was known in the middle ages as a 'flowery mead' (Plate 9), and both Basil and Augustine described how the variety and beauty of the animals in the Garden gave man a lot of pleasure (Plate 10).[6] But the fact that other authorities, Damascenus and Thomas Aquinas among them, doubted whether, after the ceremony of their naming (Plate 11), animals found a place there, shows how many crucial questions were left unanswered. Did the Garden of Eden take the form of what we would nowadays call a garden or a park? Was it formally or informally laid out? And how hard was Adam expected to labour when he was told 'to dress it and to keep it'?[7]

Even more important, though, than these somewhat technical problems of interpretation, the main thrust of the Mosaic narrative is directed, not towards the Garden, but towards the events that took place in it – towards the Fall, and men seeking to re-create the Garden of Eden had not only to ask themselves what the original had been like, but why it had been forfeited. This, in turn, was not an easy question to answer. There are two accounts of the creation in the first two books of *Genesis*: they are inconsistent, and there are a great many questions which are not resolved by either. In chapter I man and woman appear to have been created simultaneously (with egalitarian implications), while in chapter II man was created first, before falling into a deep sleep, and waking up to find that he had lost

12

a rib and gained a wife (Plate 12).[8] The difference between the two accounts could be resolved by saying that chapter I referred, in a general way, as Plato might have done, to the idea of man, and that the second chapter supplied an account of the actual processes by which the first man and the first woman were made. But this opened the door for the commentator to interpret man's original condition as having involved a union of both male and female natures, and to argue that man's unhappiness began with the creation of woman and the division between the sexes. This 'hermetic' doctrine then provided an account of the collapse of man's condition different from that given in chapter III with its description of the serpent tempting Eve with the fruit of the tree of knowledge of good and evil. This second, and more widely accepted account was itself, in turn, capable of being interpreted in several different ways. Some laid the stress upon Eve's eating the apple, and attributed the Fall, literally, to greediness:[9] others, picking upon the consequence, which was that Adam and Eve knew that they were naked, interpreted Eve's handing the apple to Adam as an allegory for the commencement of sexual relations.[10] A third group, more inclined to see hunger, gluttony and lust as consequences rather than causes of the Fall, pointed to the name of the tree, and interpreted the story in terms of man's trying to rise above his station – the creature, already endowed by a generous God with wisdom *(sapientia)* aspiring to the knowledge *(scientia)* of the creator.[11] A fourth group – a majority, perhaps – interpreted the Fall, quite simply, in terms of disobedience.[12] In due course the church, without formally deciding between the four, impartially recommended fasting, chastity, humility, and obedience to its members.

Whatever form the Fall took, *Genesis* leaves us in no doubt that Adam and Eve were expelled from the garden of Eden, and driven forth (Plate 13), Adam to win

12. (facing page) Successive scenes including the creation of Adam outside, and of Eve inside the Garden.
> G. B. Andreini *L'Adamo, sacra rapresentatione,* 1617.

13. The serpent looks up at his achievement as Adam and Eve, who now walk out of step, are driven from the Garden by the Angel with the sword.
> J. J. du Pré *Heures à l'usage de Rome,* 1488.

14. Eve hands the forbidden fruit to Adam and the trees begin to wither.
G. B. Andreini *L'Adamo, sacra rapresentatione*, 1617.

their daily bread from the cursed soil, in competition with thorns and thistles, and
Eve to spin and to bring forth children in pain and sorrow.[13] But the principal
penalty which had been threatened for disobedience ('in the day that thou eatest
thereof thou shalt surely die'),[14] was not, in fact, exacted immediately and in full,
for Adam and Eve and their descendants, the Patriarchs, are recorded as having
lived to be many hundreds of years old. A thousand years elapsed before God
resolved to send the Flood to destroy his creation, and even then, repenting of his
intention, he allowed Noah to build an ark – in which the creatures once again
lived at peace – and escape divine retribution. It was not until Noah's children
repopulated the world that the life-expectancy of man was reduced to seventy
years.

The period between Adam and Noah presents many problems, and the Christian churches never really made up their minds whether it was as a consequence of
the Fall or of the Flood that the seasons' difference (Plate 14) and the changing
length of day were substituted for the perpetual Spring equinox, that the animals
began to injure each other, the lion to hunt the deer and the hawk to prey upon the
lark, and man became carnivorous (Plate 15). Thousands of years later, perhaps it
scarcely mattered. Whenever Adam's descendants did have time to look up from
the fields, and Eve's from the cradle, they were inclined to agree with St Paul that

the whole creation groaned in travail.[15] The earth was a heap of ruins, and living things were poisoned in their natures. The level ground had been folded by earthquakes into horrid mountains and hideous deeps, and the temperate lands were now hemmed in between the frozen wastes of the North and the torrid zone to the South. Gentle dews and fertilising streams which never failed had been replaced by alternate torrents and droughts. The waterlogged clays and barren sands, the burning limes and sour peats into which the once fruitful soil had now been separated, seemed simultaneously both sterile and to support an unceasing succession of weeds. Males, separated from the anti-tumescent plants that some authorities believed to have existed in the Garden of Eden, found themselves embarrassed by an unruly member.[16] Men and women were the victims of irrational affections they could not control, and following the pains of childbirth, women found themselves burdened with young which were helpless and de-pendent for years on end, when, without the Fall, it was widely believed, children would have been able both to walk and to talk from birth.[17] Above all, men and women found themselves, at every stage of their lives, subject both to painful and crippling diseases and mental anxieties, which seemed often to bring on, and always to foreshadow, their deaths. Fall or Flood, these were among the principal evils which gardeners striving to re-create the antediluvian and prelapsarian past would seek to overcome (Colour Plate II).

15. On land, in the water, and in the air, the animals begin to rend each other.
 G. B. Andreini *L'Adamo, sacra rapresentatione*, 1617.

CHAPTER II

PARADISE

Numerous as are the difficulties of forming a conception of the Garden of Eden from the account given in *Genesis*, they are exceeded by the complications with which the topic was subsequently surrounded. In the first place, the one garden became two, for the Christian belief that the moral – though not the physical – consequences of the Fall and of sin had been overcome through the Passion necessitated a re-appraisal of the life that had been led in the Garden of Eden. Adam and Eve, it was now often said, had led lives of happy innocence in the Garden of Eden, but not of moral maturity. Basil condemned Adam's original condition as apathy, while Lactantius, Ambrose and other early Christian Fathers distinguished between innocence and virtue, and took the view that there could have been no virtue without sin.[1] This doctrine of *felix culpa*, rationalising the Fall through the Redemption (Plates 16, 17), in turn led on gradually to the conclusion that God, while ejecting Adam from the Garden of Eden, had, in his mercy, been preparing an even better home, or Paradise, as it became known from the Persian, for him in heaven.

The existence of this heavenly Paradise did not, however, mean that the Garden of Eden dropped out of the scheme of things. In the Jewish rabbinical tradition the Garden was the blessed part of Sheol where the just awaited the resurrection, and in the alternative apocalyptic tradition it was to become the abode of the blessed dead after their resurrection.[2] The early Christians adopted and developed both ideas, holding that the faithful dead were received into an earthly Paradise to await the second coming, and that after the day of judgement they would be taken up to enter a heavenly Paradise, where they would dwell with Christ for evermore. This earthly Paradise, which thus became a 'type' or prefiguration of the heavenly one, was still universally believed to be the Garden of Eden. The Garden had not been destroyed: the Jews believed that the olive branch brought to Noah by the dove as the waters of the Flood receded, had come from the Garden of Eden. Enoch, who is not recorded in *Genesis* as having died, but as having been spirited away ('and he was not; for God took him'), Elijah, who was said to have gone up 'by a whirlwind into heaven', and even the ten lost tribes of Israel were thought to dwell there yet.[3] To these residents there were added, in Christian times, in rapid succession, the repentant thief, the souls of the just led by Adam himself, whom Christ had released from bondage on the day of the crucifixion, and St John.[4] St Paul was believed to have been transported there when he was intercepted by Christ on the road to Damascus,[5] and the gate of the Garden, so the doctrine ran, stood open to Christian martyrs at all times.[6]

The judaeo-christian concepts of the Garden of Eden and of the earthly and heavenly Paradises had to spread, if they were to make any headway at all, through the Roman world. Greek and Roman literature were full of examples of a

16. (left) The rationalisation of the Fall and the Expulsion through the Crucifixion, Entombment, Resurrection and Ascension of Jesus Christ.
J. Fletcher *The Historie of the Perfect-Cursed-Blessed Man*, 1628.

17. (right) The rationalisation of the Fall: while the Devil celebrates his triumph in the original Garden of Eden, Christ leads his faithful spouse into the new Garden of the Church. This engraving from I. David *Paradisus sponsi et sponsae*, 1618, was also employed by C. Stengel in *Hortensius, et Dea Flora, cum Pomona historice, tropologice, et anagogice descripti*, 1647.

belief in an original golden age, and the subsequent degeneration of the world, as Plato described it, through successive ages of silver, bronze and iron, which might have done much either to challenge or to diversify and enrich the Mosaic conception of the Garden of Eden (Plate 18). The Greeks distinguished between Arcadia, a primitive rather than an innocent place, where the inhabitants were notorious for their rugged virtue and their ignorance, their rustic hospitality, their skill in singing, and their low standard of living, and Elysium, the land of the dead, where, according to Homer, the shades of those whom the gods favoured lived in perfect happiness on the banks of the Oceanus river at the end of the earth. But Pindar made the entrance ticket to Elysium a good life, and Virgil converted Arcadia into a land of luxuriant vegetation and perpetual Spring, so that these concepts were fairly readily assimilated. Ovid's description of the golden age included the familiar season of everlasting Spring, when 'the earth itself, . . . untouched by the hoe, unfurrowed by any share, produced all things spon-

19

18. Hercules clubs the dragon guarding the entrance to the Garden of the Hesperides. This story of the golden apples was one of several in classical literature which both resembled and differed from the Biblical account of Adam and Eve in the Garden of Eden.

R. Dodoens *Histoire des plantes*, translated by C. de L'Écluse, 1557.

taneously', and the political implications of his ideal world where 'men of their own accord, without threat of punishment, without laws, maintained good faith and did what was right',[7] was compatible with St Augustine's view that the punishments of the law and the coercive powers of rulers were the consequence of sin.

But Ovid's strictures upon private property, commerce, and navigation, appealed more to the primitive church of the Apostles than they did to the post-Constantinian established church. The manners of the young men and maidens who inhabited the classical Arcadia, dancing, and making love to the sound of music, were not the marital ones expected of Adam and Eve in the Garden of Eden; and Virgil's Elysium, which he placed underground, stood in marked contrast to the christian conception of a Paradise which lay in the open air. Ovid in his pastoral, Virgil in his agricultural, and Horace in his retirement poetry, all praised man's harmony with the seasons and with his surroundings in ways that bore little relation to judaeo-christian ideas of a ruined, poisoned and sinful world. All these differences were, however, somehow swept under the carpet. Tertullian argued that Homer's Garden of Alcinous had been suggested by the Mosaic account of the Garden of Eden, that the pagan poets had derived their ideas of Elysium from the earthly Paradise, and that they had therefore worshipped Christ even though they didn't know it.[8] Despite all his errors, Virgil was thankfully promoted to the ranks of Honorary Christians, because he had prophesied the return of the golden age, ushered in by a boy,[9] and passages taken from the classical authors were employed by Christian writers indiscriminately just as though they enjoyed biblical authority and veracity.

While classical literature succumbed to these garlands of misunderstanding, the Fathers swamped all further distinctions of language in allegory – as extensible and subtle and as difficult to escape as a net. The term Paradise (whether earthly

20

or heavenly was often left undefined) was used to describe the Church. Augustine referred to Christ as the tree of life, the saints as fruit trees, and the four gospels as the four rivers of Eden.[10] In the early church the open, pillared court (or *atrium*), where the catechumen waited before being received into the main body of the church and baptised, was called a Paradise, and in the middle ages there was often to be found, outside the church, a Paradise – an enclosed space for meditation and prayer. Benedictine monasteries all possessed Paradise gardens, and in the later Cistercian order every monk was allotted his own little plot or Paradise to look after. Throughout the middle ages, communicants were said to enter Paradise, and the soul was cultivated as a Paradise through the nurturing of the 'plant', charity, and the eradication of the 'weeds' associated with cupidity.[11] In seventeenth-century England, Sundays were spoken of as Paradises from the ravages of time (meaning that men literally didn't get any older on Sundays).[12] Every aspect of the Church's life was pictured in the same terms. Gregory of Nyssa described the life of a solitary, Jerome that of a monk in his cell, as Paradise.[13] Milton called the individual soul a Paradise, 'a paradise within thee happier far'. When the Papacy itself became corrupt, the schismatic Florentine Zion of the Dominican reformers became a Paradise,[14] and finally, in the era of the Protestant reformation, it was the sects, the congregations of the elect, who had separated themselves from the main body, which laid claim to the title.

Among the many allegorical uses of the term there was one, the custom of describing the Virgin Mary as a Paradise, in the midst of whom, in her womb, lay Christ, the tree of life, which requires a more extended treatment, because the Virgin was also identified by the Church with the *hortus conclusus*, or enclosed garden, of the *Song of Solomon* (Col. Pl. III).

> A garden enclosed is my sister, my spouse;
> A spring shut up, a fountain sealed.[15]

At first sight this connection of the Virgin Mary with the lover of the *Song of Solomon*, might, given the blatantly erotic language of the *Song*, appear strange. But Solomon himself was revered both as a kind of second Adam, on account of his wisdom, and even as a sort of prototype Christ. Solomon's marriage to the Queen of Sheba, who was black, was regarded as a type of Christ's marriage to the members of his church and to the blackness of their sins, the kisses in which the *Song* abounds were interpreted as types of the life-giving crucifixion, and the explicit language of sexual ecstasy has always held its place in the scriptures as an acceptable manner of portraying the relations between Christ and his spouse, the Church.[16]

This identification of the Garden of Eden with the earthly Paradise, of the Virgin with Paradise, and of the enclosed garden of the *Song of Solomon* with the Virgin, leading to the equation of the enclosed garden with the Garden of Eden, was to have important consequences for the history of gardening. The biblical story left it uncertain, as we have seen, whether the Garden of Eden resembled what we would call a park or a garden (Plates 19, 20). Now, in medieval England, where the fields were mainly open, and animals, if they were allowed to roam, would damage the crops, parks were established to which animals could be

19. An early sixteenth-century scene of the Fall and the Expulsion with an open Garden conceived as a flowery mead.
    J. P. Bergomensis *Suma de todas las cronicas del mundo*, 1510.

20. (left) An early seventeenth-century view of the Expulsion reduced to the simplest possible representation of a minute, enclosed Garden.
    I. David *Duodecim specula*, 1610.

21. (right) The *hortus conclusus* of the *Song of Solomon* into which nothing but the Spirit of God could enter.
    H. Hawkins *Partheneia sacra*, 1633.

confined, and gardens were enclosed in order to keep wild animals out. Except, perhaps, for Woodstock, where Henry I kept lions, lynxes, leopards, camels, and a porcupine sent from Montpellier,[17] the parks served as private hunting grounds, and at this period the Edenic ideal of peace among the animals was not attached to them. The enclosed garden behind its hedge of thorns, its wattle fence, its paling, or, if it was to be really secure, its walls, was then left with an almost undisputed claim to represent the Garden of Eden. All the ideal qualities associated with the Virgin Mary and with the earthly Paradise thus came to be identified with the small, contemporary, enclosed garden from which the animals were excluded altogether.

The enclosed garden thus became 'a secret place, enclosing within it the mysteries of the Old and New Testaments'[18] (Plate 21), and everything in this garden was then, in its turn, enveloped in allegory. Each individual flower illustrated some aspect of the christian faith, reminding the observer either of some simple virtue, or of some more sophisticated theological truth. Thus the rose, whose bud opened and whose blossom fell in a single day, put one in mind of the modesty of the Virgin, while the lily, with its white flowers, suggested her purity, and was often included in representations of the Annunciation and of the Assumption. But at the same time the rose, a flower among thorns, stood for the providential miracle of Mary's having sprung from and grown up among the Jews, who rejected Jesus. The iris, universally recognised as the *fleur de lys* and the symbol of the Kings of France, also served to illuminate Christ's descent from the royal line of David, and to conjure up the image of Christ the King. The triple lobes of the leaves of the strawberry put one in mind of the mysteries of the Trinity, and any number of plants, like St John's wort, which derived their names from the names of the Saints upon whose days their flowers were expected to open, were thought to possess either the courageous or the healing properties of their patrons. The whole garden served as a kind of surrogate Bible.[19] It was rich in allegory, and the allegory extended beyond the flowers to the features. The straight raked paths

22. When Christ showed himself for the first time, as the new Adam, to Mary Magdalene, after the Resurrection, he appeared in Adam's prelapsarian condition as a gardener.

I. David *Duodecim specula*, 1610.

and alleys reflected the conduct to be expected of a christian, the summerhouse served as a garden chapel for reflection, while the shade of a tree gave shelter both from heat and from the wrath of God, and the enclosed garden itself, like the walls of a church, offered a refuge from the deformed world outside (Plate 22).

But there was one other tradition in Christian thought which remained to challenge the received association of Paradise with the *hortus conclusus*. The wilderness might appear, simply, as the postlapsarian world in which the garden was set: the garden, as a refuge from it. But the Old Testament supported an alternative and radically different interpretation. Man had dwelt but a short space of time in the garden, and the garden had rapidly become the scene of divine retribution. It was not, after all, in the garden, but the wilderness that man had won his way back to God's favour. When Moses led the children of Israel out of their bondage in Egypt, they wandered through the desert for forty years before reaching the land of Canaan. In those forty years God fed them by his providence, gave them the Law, and made a covenant with his people. *Exodus* is the record of a whole people being educated through tribulation and privation to obedience. Time and again thereafter the prophets returned to 'the Sinai experience' for their inspiration.[20] Elijah withdrew to the wilderness and cave of Horeb, and Elisha to that of Jordan.[21] The theme can be traced right through the prophetic books from Jeremiah, through Ezekiel, to Hosea and Amos. Isaiah's reference to 'the voice of him that crieth in the wilderness',[22] and his instruction to the people of Israel that they should 'make straight in the desert a highway for our God', were taken to refer to the life of John the Baptist,[23] and Jesus himself, immediately after his baptism by John was taken away into the wilderness for forty days and forty nights to be tempted by the devil. It was not in the luxuriant foliage of a garden, leading an easy life among the fruit, that wisdom and virtue were to be found. God lay closest to the simple life, even to the life of self-deprivation, and there was more virtue in a diet of locusts and wild honey than there was in all the luscious fruits of the Orient. This development of the belief in the wilderness as a place of trial and expurgation, of refuge and contemplation, the scene of covenanted bliss, led to the expression of a bewildering variety of contrasts and similarities between the Garden and the Wilderness. At times the challenge lay in creating a Garden in the Wilderness, and the terms employed are those of the pioneer. But at other times the Wilderness itself is considered to be the real Paradise, and the language employed is that of the Puritan. The words Paradise and Wilderness, which start out as clearly defined opposites, then change places. The Garden, or Paradise, with its softness, ease, and lack of exertion, saps the fibre and becomes a moral wilderness, while the desert with its hardships, is the high road to religious experience, and is itself a Paradise.

This Wilderness of the Jews, which was a desert, was readily transmuted, in medieval Europe, where there were few deserts, into a forest. When we hear the life of a solitary hermit, or the situation of a monastery, referred to as a Paradise, we have to remember to inquire whether it may be a desert or a forest that is meant. These complications, too, affected the history of gardening, and in the Christian era it was the Saints, the hermits and the monks, the votaries of a wilderness environment, who inherited and became associated with the concept of

24

23. St Bernard contemplating God in the forest among the animals.
I. David *Duodecim specula*, 1610.

peace among and with the wild animals. Hosea had looked forward to God's entering into a covenant with the beasts of the field, the birds of the air and the reptiles of the ground, in order to secure the peace of the elect.[24] The disciple John is reputed to have anticipated a time when the creatures would once again be 'subject unto men with all obedience', and the gospel of Thomas preserves the idea of an evangelistic mission to all creatures.[25] This tradition passed straight to the Saints, the hermits and the monks, and perhaps also into the vernacular,[26] to all those who were, by location in the desert or forest, living in daily contact with wild animals. All the way from Androcles to Francis of Assisi, the mark of a holy man was his ability to charm a ferocious beast (Plate 23). The Palestinian Abbot Gerasimus is said to have relieved a lion by removing a thorn from its paw. The lion subsequently served the abbot, and when he died, collapsed over his grave and died also. 'All this was done', we are told, 'because God wished . . . to show how the animals were subject to Adam before his disobedience and his expulsion from the Paradise of delights.'[27] It was Francis of Assisi himself who brought the animals into the traditional representation of the Nativity, and it seems likely that it was the monastic orders which still preserved the ideal of peace with the animals – if it was preserved anywhere – in England until the Reformation.

At all events, the two ideals of the Garden of Eden had become separated. The perpetual Spring, which was in this period an allegory for the life and claims of the Church, was identified with the small enclosed garden from which animals were excluded, while the demands of the animals upon man's consideration met with a confused response. Friendly relations were being confined to tame animals and to the realm of pastoral poetry. Wild animals were being herded into parks and might even be regarded, as they were in the Eastern Church, as incurably depraved, and as instruments of the devil. There are no animals in Dante's description of the Garden of Eden, and Dante himself was assailed by a spotted panther, a lion and a she-wolf when he emerged from the gloomy forest where the *Divine Comedy* begins, with his eyes fixed on the summit lit by the gospel's ray.[28] Attitudes towards the animals were in a state of flux towards the end of the middle ages, and when Mantuan describes an earthly Paradise where there are no lions, no serpents and no scorpions, we begin to suspect that, behind the traditional language, what is meant is not an ideal world in which lions do not eat, serpents beguile and scorpions sting human beings, but one from which these agents of the devil have been banned, or even, perhaps exterminated.[29]

CHAPTER III

THE INDIES

RELIGIOUS and gardening allegory were interwoven, and some of the early Christian fathers professed to believe that the whole story of the Garden of Eden was an allegory. Philo held that it detracted from the grandeur of God to suppose that he planted a garden, and Origen scoffed at the whole idea, if taken literally, but added that it made sense if the four rivers were interpreted as the cardinal virtues, Justice, Temperance, Fortitude and Prudence. Ambrose was not so sure, but still preferred an allegorical interpretation in which Paradise was the soul or mind, Adam understanding, Eve the senses, and the serpent delectation.[1] But these views never commanded universal assent. Augustine wavered, and in treating of the question whether the story of the Garden of Eden was to be understood allegorically or literally came to the conclusion that it was both, adding that speculation about its whereabouts ought not to be made a matter of faith. Thomas Aquinas, who was inclined to identify the earthly paradise with the church militant and the heavenly paradise with the church triumphant, was still prepared to concede that there might yet be such a place as the Garden of Eden.[2]

Those who took the story literally can be divided into two groups. The first believed that the Garden had existed and that it had been swept away by the Flood and they, in turn, were of two kinds. Some, who were impressed by the symmetry of the Christian religion, and by the way in which Christ had atoned for the Fall, were inclined like Henry Hare, the second Lord Coleraine, to believe that the crucifixion had taken place upon the exact spot where Adam had sinned.[3] This idea found expression in the legend of Seth, who was credited with having made his way back to the Garden, where he begged seeds from the tree of life off the angel on sentry duty at the gate (Plate 24), which he planted in the mouth of his father, Adam, when he died. From these seeds, the cedars of Lebanon, were, with a pleasing sense of historical continuity, thought to have been derived, and from the wood of one of the cedars of Lebanon felled for the construction of Solomon's Temple, after many miraculous adventures, the cross itself was supposed to have been hewn.[4] Others, including the Manichees and Hugo Victorinus, identified the four rivers of the Genesis account with the Euphrates, the Tigris, the Nile and the Ganges, concluded therefore that the Garden of Eden had once covered the whole world, and argued that no true conception of its antediluvian and prelapsarian condition could now be formed from the ruins by which men were surrounded.[5]

The second group believed that the Garden had survived the Flood, and of these, too, there were two kinds. A few, like Moses Bar-Cepha, quite simply accepted that Paradise lay at a great distance separated from the inhabited world by seas or mountains which men could not cross,[6] or like Dante, that it lay somewhere in a southern hemisphere, on the other side of the torrid zone (the flaming sword which turned every way in the hand of the cherubim guarding

24. Seth, who has retraced his parents' footsteps, begs seeds of the tree of life from the Angel on guard at the gate of the Garden of Eden.
   The Buke of John Maundevill, early fifteenth century.

the gate), or even, like Martianus Capella and Adam of Bremen, that it lay beyond the polar regions.[7] Many more, starting from the view that the world had been damaged, but not totally devastated, by the Fall and Flood, looked for clues as to its whereabouts in almost any part of the world where the days and nights were the same length (a point of particular interest to Tertullian, Bonaventure and Durandus),[8] the climate was temperate, and the vegetation resembled what might be expected from a perpetual Spring, which was the case in Ceylon, the choice of Odoardus Barbosa.[9] Christian writers laid great stress upon the smell of the Garden of Eden, and upon dying in the odour of sanctity. The deliciously scented breezes noticed by sailors in the south Arabian seas lent credence to the contention that somewhere, nearby, lay the long lost and forbidden Garden from which our ancestors had been expelled, and in due course Milton wrote of perfumes carried on gentle winds which

> whisper whence they stole
> Those balmy spoils. As when to them who sail
> Beyond the Cape of Hope, and now are past
> Mozambic, off at sea north-east winds blow
> Sabaean odours from the spicy shore
> of Arabie the blest . . .[10]

In the later middle ages it seems increasingly to have been taken for granted that the Garden of Eden had survived the Flood and that it still existed. The conviction grew therefore, that it must lie very high up, either as Strabo, Rabanus,[11] and now Alexander of Neckham argued, in the sphere of the moon[12] (which was not as absurd as it sounds, given that there were seven 'heavens' or spatial orbits round the earth, and that St Paul was believed to have been taken up into the third of them),[13] or that it lay atop a high mountain. Dante placed the Garden upon the top of a mountain on an island and added, with symmetrical gain, that the island lay at the centre of the southern hemisphere exactly opposite Jerusalem, which was almost always shown, on world maps of the thirteenth century, lying at the centre of the northern one.[14] But Dante was perhaps more typical in placing Paradise upon a mountain than he was in setting it in the southern seas, and a majority of Christian authorities seem to have held, quite simply, following the Mosaic account, that Paradise lay in the East. This was the view of Isidore of Seville, of John of Damascus, of Bede and Duns Scotus, while Strabo, who situated it in the lunar sphere, combined this view with the other.[15] Many connected the idea that Paradise lay in the East with the customary alignment of churches, and with the practice of turning to face the East in prayer.[16] Even in the sixteenth century Raleigh repeated that churches were laid out 'to the point where the sunne riseth in March, which is directly over *Paradise* (saith *Damascenus*) affirming, that we alwaies pray towards the East, as looking towards *Paradise*, whence we were cast out',[17] a view which happened, in British latitudes and longitudes at any rate to be loosely compatible with the view that churchgoers turn to face Jerusalem and the scene of the redemption.

For much of the middle ages it would have been regarded as technologically impossible to go and look for a Garden of Eden lying either beyond the great

Ocean which encircled the globe, or in the southern hemisphere on the other side of the torrid zone which Aristotle said no man could traverse. But in the great age of geographical discovery, which began following the Mahometan victory in the Crusades, this attitude became out of date. The chronicler, Joinville, had recorded how, in Egypt, nets cast into the river Nile at night, yielded cinnamon, ginger, rhubarb, and other precious spices which were believed to have floated down the river from the earthly paradise.[18] The Garden of Eden must, therefore, lie somewhere in Ethiopia, and the defeated Christians acquired an additional incentive to look for new ways of furthering their search for allies among the descendants of the Queen of Sheba in Ethiopia. Had not Caspar, one of the three kings who came to witness the birth of Christ, been an Ethiopian? Had not the Eunuch in charge of the treasure of Queen Candace of Ethiopia been converted by St Philip?[19] There, among the Ethiopians, in the legendary kingdom of Prester John, Christianity was, it was believed, already established in the rear of the infidel.

At this point fact becomes woven with fiction. Sir John Mandeville's *Travels*, written about 1370, enjoyed a wide circulation in the later middle ages, and their authenticity was taken for granted. It was somehow characteristic of the age that, although Sir John Mandeville admitted that he had not actually been to the Garden of Eden, he was believed when he said that he knew it to exist.[20] However that may be, Mandeville's imaginary journeys were soon followed by voyages of a very different kind, when the Portuguese invented the ocean-going carrack, and began to explore ever further and further down the west coast of Africa. Then it was, as Baudet has said, that the Garden of Eden was removed from a distant past to a distant present, from something remote in time to something far away, but still conceivably discoverable.[21]

The bare facts are that the Portuguese reached Madeira in 1419, the Azores in 1439, and the Cape Verdes in 1456–60. Bartholomeu Dias rounded the Cape in 1488, and began to explore what was at that time known as lower Ethiopia (South Africa). Ten years later Vasco da Gama reached India.[22] All the way down the coast, past Madeira and the Azores, the Portuguese had been drawing closer to the equinoctial. The vegetation of the islands must have come more and more closely to resemble that to be expected from a perpetual Spring, and the first boy and girl to be born on Madeira were named, appropriately enough, Adam and Eve. As the Portuguese ships reached the mouths successively of the Senegal, the Gambia, the Niger and the Zaire, hopes revived of finding an inland navigation right the way across the continent to the kingdom of Prester John. Admittedly the search for the earthly paradise is not among the factors listed by Azurava as motivating Henry the Navigator.[23] But Henry's brother Pedro was credited with having secured Prester John's permission to undertake a journey through his kingdom in search of the Garden of Eden. The story was embroidered with all the circumstantial evidence one could ask for. Pedro was said to have travelled for seventeen days on dromedaries through six hundred and eighty leagues of desert. He observed the four rivers of paradise, and when he approached the Gihon, he found linaloon oaks, associated with the tree of life, floating down the river towards him. This 'proof' that he was in the right area was confirmed when the waters of the Phison bore upon their waters parrots in their nests. Before the Fall, all the beasts, it was widely believed, had been able to talk, and the parrot with its powers of mimicry and its facility for speech was interpreted as an indubitable vestige of the Garden of Eden in the same way that an arena or an aqueduct was a relic and a reminder of ancient Rome. Unfortunately, Pedro was unable to reach the Garden of Eden because his companions refused to cross the mountain ranges which still separated the expedition from it. Strictly speaking the tale tells us nothing, for the journey was apocryphal: but the popularity of the story says much about the haze of expectation in which the seafarers of the late fifteenth century moved.[24]

It should come as no surprise, then, to find that Christopher Columbus really did believe that he had discovered the location of the earthly paradise in the new continent, which for a century or more, was to be known as the West Indies. Samuel Morison, the great American historian, denied that the search for the Garden of Eden was the primary motive in Columbus' mind, and said that all the evidence pointed to one great object of Columbus' voyage, which was 'to reach the Orient by sailing West'.[25] This may well be argued, but it does not resolve the problem of what Columbus expected to find in the East. Christian authorities had always been inclined, as we have seen, to locate the Garden of Eden in the extreme Orient, though the ancient Greeks had situated Elysium and the Fortunate Islands not where the sun rose, but where it set, in the West. The thought may well have crossed Columbus' mind that the two accounts could be reconciled if paradise lay in the extreme Orient and was to be reached by sailing west. At all events, Columbus used d'Ailly's map, which showed Eden in the East, and when he read d'Ailly's chapter on the other celebrated islands of the ocean, noted that the terrestrial paradise was not in the Canaries or the Cape Verdes. The Portuguese had not found it there.[26]

Even on his first voyage in 1492 Columbus was convinced that he had arrived back in Old Testament country. Luis de Torres, 'whom he selected . . . for a reconnaissance mission, was chosen because . . . as a converted Jew he knew Hebrew, Arabic and Chaldaic – the languages that would certainly be required in the circumstances',[27] and in describing the flora and fauna, and the manners of the inhabitants, Columbus had recourse to the language of Ovid's description of the golden age.[28] On his second voyage he enthused about the lovely weather he had encountered in the West Indies, and gave his first hint, Morison says, that there was located the terrestrial paradise spoken of by 'sacred theologians and wise philosophers',[29] and on his third voyage, when he reached what is now South America, he became convinced that he had reached the region of the earthly paradise.

In a letter to the Spanish sovereigns who employed him Columbus rehearsed the whole case. Medieval map makers had placed the original home of Adam and Eve in the kingdom of Prester John, or in other parts of Africa or India, where nobody had yet found it. But many authorities had concurred in placing it at the extremity of Asia, where the sun rose on the day of creation. This is where Columbus now believed his newly discovered West Indies to lie. The latitude of the Garden of Eden was in the southern hemisphere, just below the equator, and Columbus had come close to it, or so he thought.

31

Eden had a temperate climate; and the fleet had encountered no hot weather since escaping from the doldrums. In Eden grew every good plant and pleasant fruit; and had he not found strange but delicious fruits on the shores of Paria? 'The gold of that land is good', and had he not found the natives wearing golden ornaments? 'And a river went out of Eden to water the garden; and from thence it was parted, and became four' . . . had not his men reported four rivers at the head of the gulf? Pierre d'Ailly believed that the four rivers of Paradise were the Nile, the Euphrates, the Tigris and the Ganges; and did not the immense volume of fresh water flowing through the Gulf of Paria prove that these were they? Pierre d'Ailly said that the Terrestrial Paradise was lofty 'as if the earth there touched the moon'. And Columbus's observations of Polaris on the voyage proved that he was sailing uphill![30]

So Columbus concluded that the earth was not round, but pear shaped, or that if it was round, it resembled a ball on one part of which was placed something like a woman's breast, upon the nipple of which, below the equator, lay the Garden of Eden, which he did not, however, attempt to enter because it would require supernatural grace to do so.

I have quoted from Samuel Morison's masterly biography of Columbus at length because, using his evidence, historians have since drawn the opposite conclusion from him. It *was* reasonable, in that day and age, for Columbus to suppose he had discovered the terrestrial paradise. Admittedly, Columbus' arguments in support of his claim did not convince everyone. There were scoffers like Peter Martyr, who refused to believe it, and sceptics like Las Casas, who didn't suppose Columbus had found it, but did think it was understandable that Columbus should think he had.[31] In the forty years since Morison wrote his book it has become clear that Columbus was no freak. Other explorers also couched their first impressions of the natives of the new world in terms of the golden age and of the Garden of Eden. Alonzo da Zuarza changed Hispaniola in 1518 'into an enchanted island where the fountains play, the streams are lined with gold, and where nature yields her fruits in marvellous abundance'.[32] Geoffroy Atkinson, who examined more than five hundred travel books published in France before 1610, found that the explorers and geographers fell repeatedly into the language of the ancient concepts, like Marc Lescarbot, who praised a community of natives in new France, and their life 'of the antique golden age, which the holy Apostles wanted to restore.'[33]

Spanish and Portuguese explorers lived in almost daily expectation of discovering the Garden of Eden in central or southern America. Vespucci said that the flora he had found upon his voyages reminded him of paradise.[34] Brandonio suggested that the earthly paradise lay in the torrid lands of Brazil,[35] Vasconcelos related a debate among sailors as to whether the Garden of Eden was to be found on the banks of the river Grao-Para,[36] and Gandavo, though not entirely satisfied by what he found in Brazil, seems to have reduced his expectations of Paradise in order to accommodate the identification.[37] Even the natives there seem to have shared some of the same preoccupations as the invaders, it being recorded that Tupinamba tribes migrating to the West in 1540, were inspired by dreams of a country where old women would become young again, men would be immortal, the crops would look after themselves, and where, with a twist alien to the narrative in *Genesis*, arrows would hunt the wild animals as if by magic, unguided by human hand.[38]

The tendency to interpret the newly discovered lands in terms of the Old Testament, exemplified by Columbus' conviction that he had stumbled upon King Solomon's mines at Verragua,[39] was carried a step further by the supposition that the natives of the Americas, who were clearly neither Moors nor Negroes (the two alternative racial types already known to men in Western Europe), were the ten lost tribes of Israel.[40] But there were other and more disturbing aspects to the new world which made it more difficult to continue to interpret it in conventional terms. The discovery of so many hitherto unknown plants and animals, of plants for which there were no Biblical names, like maize, the potato, cassava, Indian beans, the tomato, tobacco, vanilla, chocolate, the pineapple, the sunflower, the nasturtium, the morning glory and the marvel of Peru, and of animals like the llama, the bison, the turkey and the guinea pig, the iguana, the toucan and the anaconda which were not recorded anywhere as having boarded the Ark led to fundamental questions about the creation.[41] Had there, perhaps, been more than one creation? Even the native inhabitants might not, after all, be the descendants of the ten lost tribes. At a very early date Acosta was worried because the Jews possessed an alphabet and the Indians had none.[42] Second impressions revealed that in one place or another Indians practised cannibalism, ritual massacre, incest and sodomy – none of which was customary among the Jews. Paracelsus suggested that the American Indians were descended from another Adam, and in the seventeenth century Isaac de la Peyrère postulated a first creation of the non-Jews, a second creation of Adam and the Jews, and a purely Palestinian Flood. Peyrère's book was burnt, but men continued to speculate upon *creatio palaeogeana* and *creatio neogeana*, and began to think in what are now called polygeneticist terms.[43] Even the passion flower, which seemed a sure mark of God's handiwork in America, was offset by the Creator's extraordinary neglect in failing to provide either wheat or wine for the celebration of mass, or olive oil for sanctuary lamps. Did God intend the newly discovered lands to be Christian or not?

The disappointment to which these problems led did not come all at once. The Portuguese continued for a century and more to look for the original Garden of Eden in Brazil. When the English began to take a part in the exploration of the new world, the reaction of the first arrivals in North Carolina, Philip Amadas and Arthur Barlow, was exactly the same as that of their Spanish and Portuguese predecessors. They reported that 'We found the people most gentle, loving and faithful, void of all guile . . . and such as lived after the manner of the Golden Age. The earth bringeth forth all things in abundance as in the first creation, without toil or labour.'[44] Thomas Hariot's *Briefe and True Report of the New Found Land of Virginia* published in 1588, described a pastoral paradise, Hakluyt still expected Eden to be discovered south of the equinoctial line,[45] and the most systematic attempt

to locate the Garden of Eden in the New World can be found in two thick volumes that Antonio de Leon Pinelo wrote in the early seventeenth century – *El paraiso en el nuevo mundo*. This baroque scholar sought to demonstrate . . . that the four great rivers of South America – the Plata, the Amazon, the Orinoco, and the Magdalena – were actually the four rivers flowing out of the Garden of Eden.[46]

But the conviction gradually spread that since the earthly paradise had not been discovered, like the phoenix and the unicorn, it was never likely to be discovered now. Nowhere is this more arrestingly put than in Raleigh's *History of the World*. Raleigh was the captain of the company to whom Philip Amadas and Arthur Barlow reported, and he subsequently made an expedition to Guiana in person. He was, therefore, in a position to know, when he declared that 'if there be any place upon the earth of that nature, beautie, and delight, that *Paradise* had, the same must be found within . . . the Tropicks' (Plate 25). As he said, he knew of no other part of the world of better, or even of equal temper. The tropics boasted goodly rivers and stately cedars, and 'so many sorts of delicate fruites, ever bearing, and at all times beautified with blossoms and fruit, both greene and ripe, as it may of all other parts bee best compared to the *Paradise* of *Eden*.' The Indies, with their 'perpetuall Spring and Summer' had the best claim to be regarded as the earthly paradise, but that did not mean that it was there. 'Lay downe by those pleasures . . . the fearefull and dangerous thunders and lightnings, the horrible and frequent earthquakes, the dangerous diseases, the multitude of venimous beasts and wormes', and there could be no doubt that the inconveniences outweighed the advantages. Then came the 'modern' twist to his sceptical conclusion: 'nature being liberall to all without labour, necessitie imposing no industrie', men themselves became corrupt, and 'idlenesse bringeth forth no other fruites than vaine thoughts and licentious pleasures'. The putative earthly paradises were '*Vitious Countries*'.[47]

The same thought seems to have affected other writers whose works could in no sense be described – as some of Raleigh's were – as atheistical. Saluste du Bartas (Plate 26), a French prototype of Milton, author of *His Divine Weekes and Workes*, and a great favourite with James VI and I, described the original Garden of Eden, and then raised the possibility that

> roaming round the earth,
> Thou finde no place that answeres now in worth
> This beauteous place.[48]

Charles Stengel, the Abbot of Anhusanum in Wurttemberg, in his *Hortensius*, published in 1647, pointed to the fact that nobody had ever yet found the earthly paradise, even though the whole earth was now known and inhabited.[49] Disappointed by their failure to discover an earthly paradise in what it was now clear was a new world, men were bound to conclude that the original Garden of Eden had, after all, been destroyed by the Flood. Milton suggested that the Garden of Eden had been moved from Mesopotamia,

> Out of his place, pushed by the horned flood,
> With all his verdure spoiled, and trees adrift
> Down the great river to the opening gulf,

where it had taken root

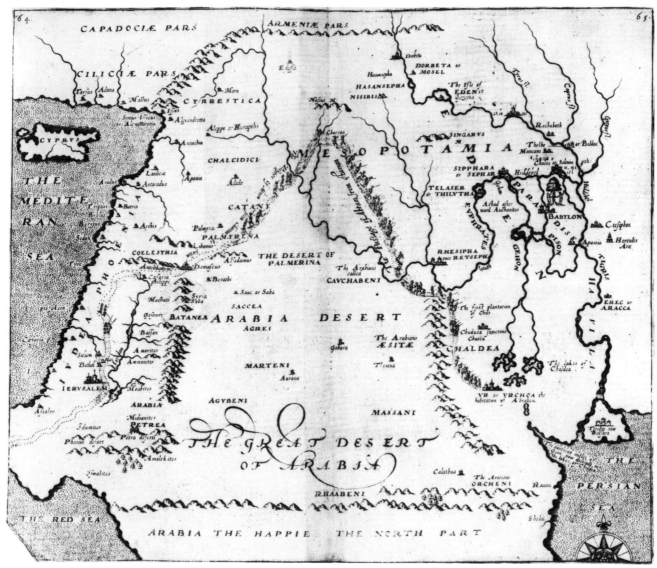

25. For all his interest in the tropics Raleigh marked Paradise, the four rivers of Paradise, and the land of Havila where there was gold, in Mesopotamia.
 The History of the World, 1614.

> an island salt and bare,
> The haunt of seals and owls, and sea-mews clang.[50]

It remained, therefore, only, as a matter of archaeological interest, to identify the site. The horizons of inquiry became once again more modest and more realistic. In the middle of the sixteenth century Augustus Steuchus, the keeper of the Vatican Library, undertook to bring the search for the Garden of Eden back to the Biblical evidence, and suggested that Paradise had lain in Babylonia, somewhere about the confluence of the Tigris and Euphrates. Franciscus Junius placed it a little further north, and his opinion was accepted by both Roman Catholics and Protestants.[51] Not everyone agreed, and the exact location of the original Garden continued to be a matter of curiosity right up to the end of the seventeenth century and beyond. Thomas Burnet reopened old controversies in his *Telluris theoria sacra*

26. (facing page) The frontispiece to Saluste du Bartas's *Works*, 1611, evokes the glory of God, and the goodness of the Creation. The sun is positioned over Adam, and the moon, the lesser luminary, over Eve, as she accepts the apple.

27. (left) In the frontispiece to his *De la situation du paradis terrestre*, 1691, P. D. Huet placed the Garden of Eden close to the shores of the Persian Gulf.

28. (right) Salomon van Til's Garden of Eden is regularly planted with trees and sited far upstream in Mesopotamia.

     Dissertationes philologico-theologicae, 1719.

published in 1681, by suggesting that the Garden of Eden had lain in the southern hemisphere, and that it was not until the Flood that the penalty of the seasons' difference was imposed. His treatment of the theme was probably more influential as a prologue to the elucidation of a science of geology, with its schools of Neptunists and Vulcanists, with their rival theories as to the agency of water and fire in laying down the different strata, than it was as an epilogue to the search for the Garden of Eden. Even though, in this latter context, Burnet's theories were at once assailed as being both erroneous and heretical, much of the urgency had gone out of the question. As Nathaniel Cotton lightheartedly said

> It puzzles much the sages' brains,
> Where Eden stood of yore
> Some place it in Arabia's plains,
> Some say it is no more.[52]

With Huet, whose *Traité de la situation du paradis terrestre* (Plate 27) was published in 1691, van Til, whose *Tractatus de situ paradisi terrestris* (Plate 28) appeared in 1719, and Delitzsch, whose *Wo lag das paradies?* became available in 1881, the argument remained firmly narrowed down to the difference between Armenia and Mesopotamia–Babylonia.

CHAPTER IV

THE BOTANIC GARDEN

THE discovery of America posed a challenge to the authority of the Bible. But the ease with which the invading European men, animals and plants advanced, and the rapidity of the collapse of the indigenous civilisations in the face first of firearms, and then of European diseases, meant that the natives were soon reduced to the role of extras upon the scene of European expansion.[1] The threat they posed to the authority of the scriptures was put out of sight, and instead it was the credibility of the classics that suffered. Aristotle had said that the equatorial zone was too hot for life. But the Portuguese had journeyed down the West and up the East coast of Africa, and both they and the Spaniards were now crossing the equator regularly upon the East and West coasts of South America respectively. In 1580 Joseph d'Acosta, sailing to the Americas under the overhead sun, felt a great cold: 'what could I else do then but laugh at Aristotle's . . . philosophie?' It was not just that Ptolemy's world map showed no trace of America; in Pliny's thirty-seven books of natural history there was no mention of the llama; and Hippocrates, Galen and Avicenna had not a word on syphilis between them. The post-Columban generation concluded that 'Ptolomeus, and others knewe not the halfe'.[2] The authority of the classics, as science, was shaken, and some men, like Hobbes, cast scorn on classical political philosophy too. As literature the ancient authors continued to hold their own, and Horace's second epode, for example, enjoyed a boom, particularly among those who were banished from court during the civil wars in France and England, and became a best seller in sixteenth- and seventeenth-century Europe.[3] Even in a field so narrowly balanced between the arts and the sciences as architecture, Vitruvius remained a source of inspiration to late Renaissance princes. But as text books to the natural world the classics could never recover, and the result was to throw men forward into new observations in what was known as the book of God's works. In every field the sixteenth and seventeenth centuries mark the beginning of modern science. Gilbert and Harvey advanced the study of magnetism and the circulation of the blood, not from ancient books, but from observation, and taught 'a new thing'.

At the beginning of the sixteenth century Vives, Rabelais and Ficino all encouraged men to study the facts of nature, and to come to know God through an acquaintance with his works.[4] As the century wore on, this approach began to appeal to all those, both Roman Catholic and Protestant, who regretted seeing Europe divided into two warring religious camps. There was good sense in this, for, as Ralph Austen put it in 1657, 'the people of God in the beginning of the world were without the Scriptures for many yeares, and they read many things in the book of the creatures.'[5] The book of God's works was older than the Bible whose interpretation caused so much controversy, and the hope was that by a common endeavour to observe nature and exchange information, scientists might recover the shattered unity of Christendom. Among those concerned to reinterpret the Fall, and justify the ways of God to men, both the Frenchman du Bartas and Milton sang the praises of the study of God's works. Du Bartas, who inquired curiously how it was that the American continent had not been known to the ancient Greeks and Romans, and speculated where the inhabitants had come from, said that he loved

> to looke on God; but in this Robe
> Of his great Works, this universall Globe,[7]

while Milton referred to Heaven

> as the book of God before thee set,
> Wherein to read his wondrous works . . .[8]

and pointed out how the marvels of this universal frame could be perceived, admittedly more dimly, in the lower works of the creation accessible to man. Throughout the seventeenth century the chorus is unmistakeable. Mildmay Fane proclaimed that nature's book was not eccentric but divine,[9] William Prynne explained how, 'If Bibles faile' each garden will descry the works of God to us,[10] and 'Anthracius Botanophilus' spoke of kind nature always having 'held forth her Book'.[11] All, no doubt, would have agreed with du Bartas that

> The World's a Booke in *Folio*, printed all
> With God's great Workes in Letters Capitall.[12]

The thought behind the idea of reading in the book of God's works involved some modification of the view that all nature had been poisoned at the Fall. If God had revealed an aspect of himself in each plant and animal that he created, the creatures could not be wholly depraved, and with the discovery of America the idea grew up that what had happened at the Fall was not so much that nature had been poisoned, but that it had been scattered. It was as though the creation was a jig-saw puzzle. In the Garden of Eden, Adam and Eve had been introduced to the completed picture. When they sinned, God had put some of the pieces away in a cupboard – an American cupboard – to be released when mankind improved, or he saw fit. Columbus had not found the earthly paradise, but he had revealed a fourth continent (few people at this time supposed there was a fifth in Australia). The result was that the defects in what had hitherto passed as the world picture had now become apparent. In the later middle ages men's comprehension of the world was expressed in what are known as the T in O maps (Plate 29), where the earth appeared as a disc or O, while the Mediterranean was the downstroke, and the Black Sea and the Red Sea the arms of the T. The three continents were then placed in the three segments, Asia, the large continent in the upper segment, Europe and Africa in the smaller segments at the bottom.[13] The Garden of Eden or earthly paradise itself was frequently placed at the very top, in the extreme orient, and was sometimes shown on an island lying beyond the waters of the Oceanus river.

Three had been a 'good' number, as anything corresponding to the Trinity must be, and when the three Kings (of medieval tradition) appeared in the

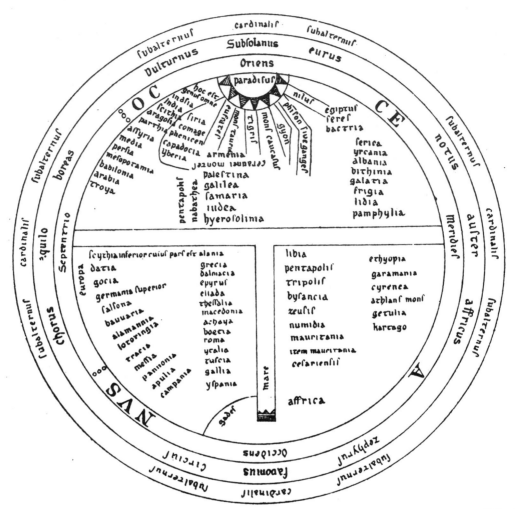

29. The T in O arrangement of the thirteenth-century world map of William of Tripoli. Geography, like history, was interpreted in terms of the Crucifixion, bringing Palestine to the centre of the world and to the top of the Cross, while the Garden of Eden, or Paradise, was shown in the extreme Orient where the sun rose on the day of creation.

nativity play they were generally taken to represent the three continents,[14] but popular language had always referred to the four quarters of the globe, and ancient Persian gardens had actually been laid out to represent them. The conventional garden of western Europe was a square enclosure divided by paths into four quarters with a fountain at the centre, and gardeners were perhaps particularly receptive to the idea that these might correspond to the four continents. At all events this theme of the four continents crops up time and time again after the discovery of America, though nobody seems to have tried to include a fourth king in the nativity play. It can be seen at its most striking in the frontispieces to du Bartas's *His Divine Weekes and Workes* (translated in 1605), to Parkinson's *Theatrum Botanicum* (1640) (Col. Pl. IV), and to Heylyn's *Cosmographie* (1652) (Plate 30). But it also appears in other forms of literature and among the

30. (facing page) The frontispiece to Heylyn's *Cosmographie*, 1652, illustrates the current preoccupation with the theme of the four continents.

representational arts. No author was more aware of the significance of the discovery of America than Abraham Cowley. In his *Plantarum*, written in Latin about 1662, and first translated into English in 1721, Cowley speculates on the reasons why God, in his providence, had kept the Indies hidden for so long, waxes lyrical about the light that Columbus has opened to mankind, recognises that, contrary to what Ovid had said, navigation has a role to play in the realisation of the Golden Age, gives full weight to the products of this new world that 'for long Ages lay conceal'd', and to his great credit does attempt to imagine what the Indians thought of the Spaniards.[15] The same basic theme of the four continents appears in the procession of cavalry costumed to represent Europe, Asia, Africa and America at the fête in the Boboli Gardens in Florence in 1661,[16] among the statues in the gardens at Versailles, and in Jacob van Meurs' emblematic painting, with its European horse, its Asian camel, its African leopard and its American parrot, of the tribute reaching Amsterdam[17] (Plate 31). Finally it is worth noticing that in his *History of the Wonderful Things of Nature* John Jonstone claimed to be able to recognise the marks of the four continents in every man.

As Cowley said, with the discovery of America, new forms of animals, new plants, and new fruits had surprised men's sight. Now that God had revealed the hitherto withheld part of the creation, men could go a long way towards re-creating the Garden of Eden by gathering the scattered pieces of the jigsaw together in one place into an epitome or encyclopaedia of creation, just like the first Garden of Eden had been. The collection should include, as Milton said,

> Whatever Earth all-bearing mother yields
> In India east or west, or middle shore.[18]

This, the idea of searching 'the *Indies* for their Balm and Spice' in order to 'Rifle the treasure of old *Paradise*',[19] was it seems, the ultimate motivation behind the creation of Botanic Gardens in the sixteenth and seventeenth centuries. There had been Botanic Gardens before, but this particular approach could not have succeeded before the discovery of America, and a clear line was drawn between the earlier and the modern Botanic Gardens. Aristotle had kept a Botanic Garden, with Theophrastus as curator, Burckhardt says that a Botanic Garden and a menagerie were part of the style and standard of living expected of a fourteenth- or fifteenth-century Renaissance prince,[20] and the Spaniards were impressed by the botanical and zoological compounds which they found in Montezuma's capital, Mexico City. But none of these collections could, at the time it was formed, have been complete, and the first modern Botanic Garden was that founded at Padua (Plate 32), the university city of Venice with its interest in the geographical discoveries, in 1545. Most notable among the many others which followed, were those at Leyden (Plate 33), founded in 1587, and two gardens at Paris – Robin's, and the *Jardin du Roi* founded in 1626. In England, John Gerard attempted to enlist the support of the Lord Treasurer, Burghley, to persuade the university authorities to establish a Botanic Garden in Cambridge.[21] But Burghley died in 1589, nothing came of the project, and the first Botanic Garden was that founded by the Earl of Danby, at Oxford, in 1621.

31. Jacob van Meurs's emblematic engraving of the tribute of the four continents reaching Amsterdam, 1663. Copyright the Rijksmuseum, Amsterdam.

Historische
Beschryvinghe
van
AMSTERDAM.

T'AMSTERDAM,
By Iacob van Meurs Plaatfnyder en Boeckverkooper inde Nieuwe ftraet, In
de Stadt Meurs. Anno 1663.

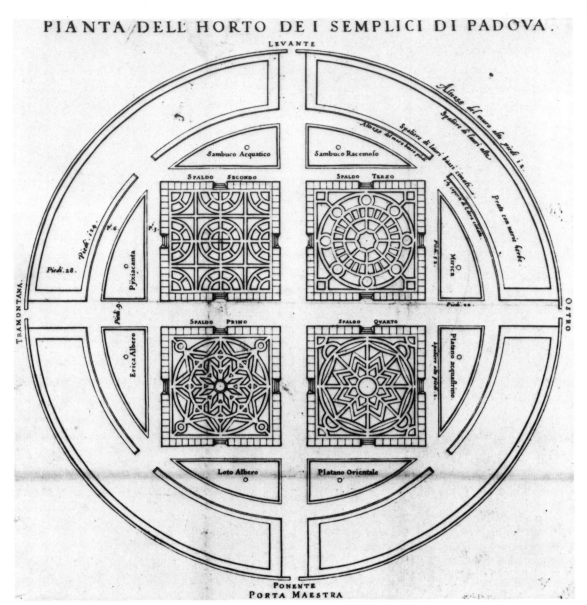

32. The plan of the Garden at Padua.
G. Porro *L'horto de i semplici di Padova*, 1591.

In the guides to these great gardens the theme is always the same – it is that of gathering the plants together from all over the world. At Padua, Porro said they were collecting the whole world in a chamber; at Leyden, Hermann described the plants as being derived from the whole world and from either Indies; at Paris, Robin was said somewhat fulsomely to have flourishing in his garden everything that a bountiful earth offered, while the *Jardin du Roi* featured plants from the frozen arctic and from scorching Africa, from the banks of the Ganges and the shores of the Cape.[22] The catalogue of the plants in the *Jardin du Roi*, compiled in 1656, ends with a list of plants brought back from America. In Oxford, whither specimens were brought, as Thomas Baskerville said, from 'the remote Quarters of

44

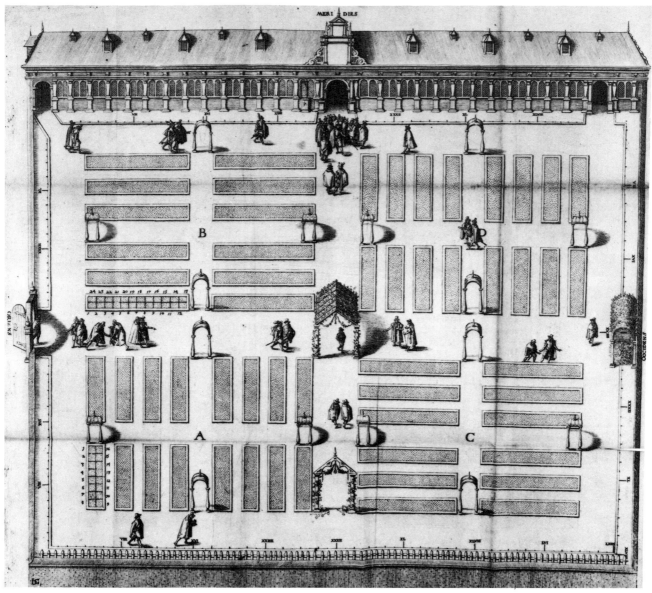

33. The Garden at Leyden.
P. Paaw *Hortus publicus academiae Lugdunum-Batavae*, 1601.

the World',[23] the garden contained plants 'comprehended as in an Epitome'.[24]
The catalogue of 1658 listed two thousand plants, of which, it was said, no more
than six hundred were English, and in 1713 'Vertumnus' described how there was
nothing

> in *Africk, Asia,* shoots
> From Seeds, from Layers, Grafts, or Roots;
> At both the *Indies,* both the *Poles,*

but it lay open to inspection in 'the Gardens' Garden of the land'.[25]
The Abbot Stengel, writing in 1647, picked out as models of what an Edenic

garden should be, Padua and Leyden, where plants had been brought in from all over the world,[26] and deeper still in central Europe, in Bohemia, 'observers marvelled at the profusion of strange plants and creatures' which could be seen around the Court of Rudolf II.[27] Here, in these Botanic Gardens, as Edmund Spenser said in his description of the Garden of Adonis,

> every sort is in a sundry bed
> set by itself, and rankt in comely rew.[28]

In these epitomes of the world, the four quarters of the garden came to represent the four continents, and there is also a little evidence, though not as much as one would expect, to show that attempts were made to plant them out accordingly. In his account of the garden at Padua, Porro drew attention to the fact that the plants which came from the East (mentioning the cedar, the laurel and the myrtle among others) were to be planted on the eastern side of the garden. Much later there is a confused record of geographical planting in the garden at Oxford, and in 1758 John Hill, in his *Idea of a Botanical Garden in England*, suggested including among other features demonstrating the various ways of ordering the plants, four quarters. These would be 'appropriated to the four great regions of the earth, and defined for the reception separately of European, African, American, and Asiatic plants.' It was a scheme which had already, he supposed, been put into practice at Goodwood, and he was confident that there were enough plants coming from each of the four continents to fill each of the four quarters. Entering these would then 'be like visiting those great divisions of the earth in succession'.[29]

This theme, the collection of plants from all over the globe, and the realisation of the world in a chamber, was not confined to Botanic Gardens. It appears, too, in plant books like Parkinson's *Theatrum Botanicum* of 1640, 'an herball of large extent . . . encreased by the access of many hundreds of new, rare and strange plants from all parts of the world.' William Broad commended Parkinson's work on the ground that everything the world offered was to be found here, illustrated, in this garden of a book, and in the frontispiece each of the four continents was associated with what were believed to be its own peculiar animals and plants – Africa being represented by the zebra, and America by the llama, the cactus and the passion flower among others. Nor, among books, was the theme confined to manuals. Hawkins's devotional garden, in his *Partheneia Sacra*, contained 'a Monopolie of all the pleasures and delights that are on earth, amassed together',[30] and nowhere was the theme more fully developed than in Cowley's *Plantarum*, in books II and V where he described two councils or contests among the plants. The first was set in the Botanic Garden in Oxford, where, after the gate had been shut for the night and the curator had gone to bed, the medicinal plants debated which of them was of most value, and the second took place on a new Fortunate Island in the Atlantic, where Cowley imagined that Pomona, the goddess of fruit, forsaking the now despised Garden of Alcinous, had taken her seat.

> And wisely too, that Plants of ev'ry Sort
> May from *both* Worlds repair to fill her Court.[31]

Here the Eur-Afro-Asians disputed the juicy fruit championship of the world with the Americans.

The Botanic Gardens, which are the most perfect examples of the attempt to collect the whole world in a chamber, may be said to have lain at the academic end of the gardening spectrum. But it would be a mistake to think that the same theme did not contribute inspiration to other gardens of the period – not least the gardens of Renaissance princes, where trees and bushes might be laid out in rings to represent the orbits of the planets, and waterworks installed to illustrate the laws of hydraulics and mechanics – all revelatory of the laws of nature and the works of God.[32] No man could have been more representative of the English gardeners of the time, or more respected by their successors, than John Evelyn. The owner of Wotton and of Sayes Court was well acquainted with the gardens of the great, and had travelled extensively through the Low Countries, France and Italy, inspecting royal, noble and botanic gardens wherever he went. When he returned to England to re-make the garden at Wotton, under the escarpment of the North Downs in Surrey, and to write about gardening, his mind was filled with ideas for the re-creation of Paradise. In the *Kalendarium* he said that our gardens should be made 'as near as we can contrive them' to resemble the Garden of Eden,[33] and in his latest, most ambitious and compendious work the 'Elysium Britannicum', which never got beyond manuscript, and of which one volume only, out of two, remains, he set out to describe his ideal garden.

No one should be misled by the word Elysium into supposing that Evelyn's patterns of thought were more classical and less biblical than those of most of his contemporaries. Evelyn came by the title towards the end of a paragraph which started with Adam, called the Protoplast, and the Garden of Eden, called the Paradise of God, continued through the identification of gardens with both 'the church Catholic on Earth' and 'the Triumphant in Heaven', and ended with Kings and Philosophers who, 'when they would frame a type of Heaven . . . describe a Garden and call it ELYSIUM'.[34] The book was written for royalty, 'Princes, noble-men and great persons who have the best opportunities and effects to make gardens of pleasure',[35] and the complete garden would have many parts. But 'a principall part' would be assigned to the philosophical garden, 'such an ample plott . . . as may suffice to comprehend the principall and most useful plants, and to be as a rich and noble Compendium of what the whole Globe of the Earth has flourishing upon her boosome.'[36] In short, Evelyn wanted to introduce into the perfect regal garden as he conceived it, a botanic garden of his own (Plate 34), and in order to show what he meant he discussed this proposed garden in the context of the Botanic Gardens at Padua and Pisa, with their eastern plants, at Leyden with its 'Indian' plants, at Montpellier with its outstanding collection of alpines (he calls them Alpeswall plants), and at Oxford. But the Botanic Garden which appealed to him most was evidently that at Paris (Plate 35). It was the largest, with its eighteen acres, and it was contrived in a different way from the others, (and, as we would now say, a more scientific one), with its several habitats, its "Groves and hills, meadow ground, and flat marshie'. This was the garden Evelyn would adapt for his Elysium or Garden of Eden, and his sketch of his proposed Botanic Garden exhibits features reminiscent both of Leyden and of Paris. There are the same little rectangular beds set out in a regular pattern as at Leyden, and there is, in addition, a mount, as at Paris. The four sides of the mount

34. John Evelyn's sketch for his philosophical garden. 'Elysium Britannicum'. Copyright the Trustees of the Will of the late Major Peter George Evelyn deceased.

were to face the four points of the compass, and it was to rise in steps, six feet at a time, to a height of seventy-two feet. The excavation of the soil would 'hollow and abate a proportionable part' round about it so that when it was finished Evelyn would have a theatre for shade-loving plants in the North, basins for marshy and aquatic plants in the East and West, and a level plain to the South where the sun loving plants would grow.[37]

One other seventeenth-century garden must be mentioned in this context, and that is the garden founded by the Dutch East India company at the Cape, not so much for what it was, but for what it was thought to be. There can be little doubt that utilitarian considerations were paramount in its construction. The Portuguese who used to sail direct to Goa, were accustomed to losing up to four hundred men out of a crew of between six and eight hundred on a single voyage.[38] When the Dutch first settled at the Cape in 1652, their object was to provide a station for their ships sailing between Holland and the East Indies. Ships reached the Cape with up to one hundred or one hundred and fifty sick,[39] and the settlement was laid out with a fort, a hospital, a church and a cemetery, and lying

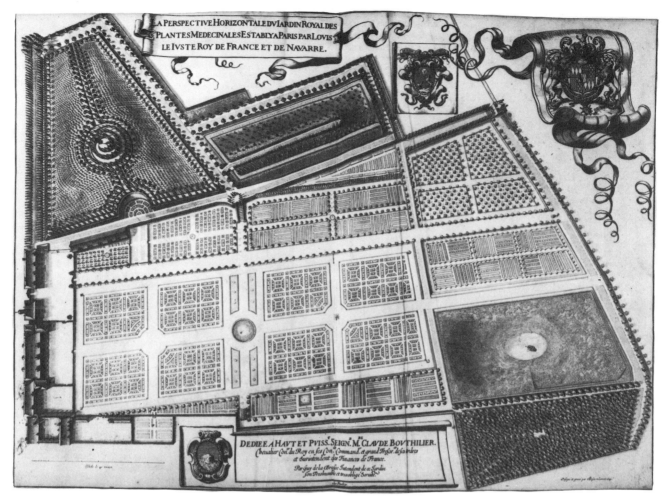

35. The *Jardin du Roi* at Paris with its different habitats. The mount can be seen in the top left-hand corner, and the lower ground lies towards the bottom right.

G. de la Brosse *Reliquae operis historici plantarum*, 1641.

next to the hospital a great garden approximately one thousand paces long by two hundred and sixty in width. From this garden the Dutch supplied fruit and vegetables to the scurvy-ridden sailors, and thus far the demands of commerce were clearly dominant. But the trade which the Dutch, with their world-wide connections, were engaged in, included plants, and the Cape was perfectly situated at the cross roads of global navigation, to act as a holding ground for the exchange and transmission of plants of all kinds between the four continents. What had begun as a vegetable garden soon became a garden of acclimatisation. Already by the time of Father Guy Tachard (his account was published in 1686) there were fruits there 'which are found in all parts of the world and have been transported to the garden', and in Kolb's account (published in 1719) he drew attention to the 'curious plants, flowers and trees which have in part been introduced from foreign countries . . . and in part taken from the wild country of this region'.[40] There was nothing, it was supposed, that would not grow there, and Sir William Temple, who had retired from London to his garden at Sheen to grow vines and apricots and to write a great book on gardening, interpreted it as a

49

36. Credulity: the tree whose leaves, when they drop, turn into birds if they fall on land, and into fishes if they fall into the water.
     C. Duret *Histoire admirable des plantes*, 1605.

37. (facing page) The Scythian lamb or borometz.
     C. Duret *Histoire admirable des plantes*, 1605.

deliberate attempt to collect the plants of the whole globe, and to exhibit them appropriately by dividing the garden into four quarters, one for the plants of each continent. As he said, 'there could not be . . . a greater thought of a gardener'.[41] Those on the spot knew better: the climate was less temperate, and the winds more boisterous than Sir William supposed. There were many plants that would not flourish, and some that could not grow there at all.[42] But Sir William was certainly not alone in the emphasis he laid upon the collection and display of plants from all four quarters of the globe, and in the early eighteenth century, the Anglo-Dutch financier George Clifford possessed four greenhouses in his garden and private zoo at the Hartekamp in the Netherlands. One house was for the plants of southern Europe, a second for the Asian, a third for the African, and a fourth for the tender American plants, and the frontispiece to Linnaeus's *Hortus Cliffortianus* shows an Arabian woman bringing a plant of coffea arabica from Asia, a negro with an aloe from Africa, and a befeathered Indian with hernandia from America.[43]

There were practical and obvious reasons why animals had to be excluded from a Botanic Garden. But the Christian universalists who thought to collect all the plants into one garden in order to see the many faces of God as manifested in his works, sought to exhibit all the animals too. As Evelyn said, 'after plants, Solomon . . . discuss'd likewise *de reptilibus*', and Mildmay Fane sang of the '*liber creaturarum*'.[44] In the sixteenth and seventeenth centuries men still thought in terms of a great chain of being (one of the Greek scientific concepts that had survived) linking everything from the rocks, which are all matter, up through the plants which have life but not locomotion, the animals which have locomotion but not

reason, and men with their dual nature, half-material half-spiritual, to angels, which are all spirit.[45] It was an essential aspect of this theory that there were no gaps in the chain, and that there is, therefore, no clear dividing line between plants and animals. Thus the link between the rocks and plants was a sponge, and the link between plants and animals that Aristotle, who was a marine biologist, would have talked about was a sea anemone, a plant with a carnivorous diet. In the later middle ages the example given might have been that of trees, growing by the banks of rivers, whose fruits, when they fell, became geese (Plate 36),[46] and in the seventeenth century one would probably have been referred to the Scythian Lambs. These creatures, which were believed to live in what was still a remote part of Asia, were said to have roots and a stem like a young tree, and the body of a lamb perched on top (Plate 37). The lamb ate the grass round its single rooted 'foot', and then died, propagating itself by seeds like other plants. Du Bartas described them as true beasts 'in Scythia bred'

> Of slender seeds, and with greene fodder fed
> Although their bodies, noses, mouthes, and eyes
> Of newly-yeand Lambs have ful the forme and guise,
> And should be very Lambs, save that for foote,
> Within the ground they fixe a living roote,
> Which at their navell growes, and dies that day
> That they have brouz'd the neighbour grasse away.[47]

Francis Bacon wrote sceptically about them in his *Sylva Sylvarum* published in

1627,[48] but two years later one appeared on the frontispiece to Parkinson's *Paradisi in Sole*, *Paradisus Terrestris*. In 1640 Parkinson offered a description in the text of *Theatrum Botanicum*,[49] and as late as 1657 John Jonstone still featured them in his *History of the Wonderful Things of Nature*.[50]

Given this continuing belief in the great chain of being it is scarcely surprising that in most of the accounts we possess of these early Botanic Gardens, mention is made of other collections, both of rocks and animals. Porro says that the buildings attached to the Botanic Garden at Padua were designed to display minerals, birds, and fishes.[51] Aldrovandi did not confine himself to plants, and is said to have opened, at Bologna, what was virtually the first museum of natural history in Europe.[52] Pawi says that the galleries along the length of the garden at Leyden (Plate 38) were constructed to house collections from both the East and West Indies.[53] In England, the Tradescants collected into their 'arke', 'beasts, fowle, fishes, serpents, and worms', as well as 'precious stones', and objects of what would now be called anthropological interest like 'the passion of Christ carved very daintily on a plumstone'.[54] Several generations after the Tradescants, Seba, a German apothecary domiciled in the Netherlands, sold his unrivalled collection of curiosities, for 15,000 gulden, to Peter the Great.[55]

This interest in rocks and stones, which was at least as old as Ernaldus, who said that he had found virtue to exist even in gems 'which are rolled about in the gravel of a stream's head-waters',[56] appears to have been carried forward in many of the great gardens of the seventeenth century, where the bottoms of artificial ponds, with their clear waters, were paved with imported coloured pebbles, and even, it is tempting to suppose, into the eighteenth century when Alexander Pope furnished his grotto under his villa at Twickenham with stones collected from all over the world.

> Lyttelton sent red spar from lead mines, Spence, pieces of lava from Mount Vesuvius . . . Sir Hans Sloane presented stones from the Giants' Causeway; the Dowager Duchess of Cleveland, clumps of amethyst and spar; Dr Borlase, the Cornish antiquary, contributed his native diamonds, ores and coloured mundic; Mr Cambridge, large pieces of gold cliff, Brazilian and Egyptian pebbles and blood stones, petrified wood and moss, fossils and snake stones, gold ore from Peruvian mines, crystals from the Hartz mountains, silver ore from Mexico, coral, copper ore, spar from Germany and Norway, curious stones from the West Indies.[57]

As for the animals, Evelyn, in his *Elysium Britannicum* included a section on Vivaries, 'greene and shady places . . . where wild beasts are kept', which were 'by the Greeks call'd Paradises'. His ideal garden would feature enclosures for animals, formed by 'wales, partitions and accommodations', such as he had seen at Brussels, and he commended the manner in which the Turks kept 'Bears, Wolves, Leopards, and even the fiercest Lyons and Panthers', though he ended rather tamely, in what is admittedly an unfinished manuscript, by suggesting that the British might run to squirrels and tortoises.[58] There was an extensive series of menageries in the gardens at Versailles, and all accounts speak of a menagerie of some sort just outside the great garden at the Cape, though there is little indication

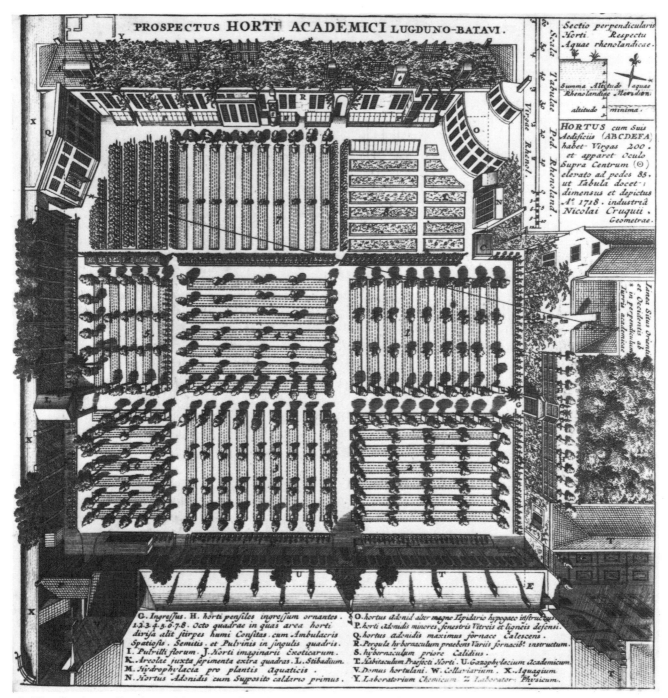

38. The Garden at Leyden, with its galleries, already greatly enlarged by the beginning of the eighteenth century. H. Boerhaave *Index alter plantarum*, 1720.

of what was kept there. In Rudbeck's time there were exotic animals kept in the Botanic Garden at Uppsala,[59] founded in 1657, and in the eighteenth century that unrivalled collector, Lady Margaret Cavendish Bentinck kept both a botanic garden and a considerable menagerie upon her husband the Duke of Portland's estate at Bulstrode in Buckinghamshire.[60]

All in all, then, it seems clear that many of those who subscribed to the Botanic

Garden ideal would have wanted to keep animals had they been able. Given their ignorance of the problems involved in moving animals, it is not surprising that some, even of the great gardeners, had to be content with representations carved in yew or box, or with models as in the grotto at the Villa Castello. Nor is it to be wondered at that, in the face of all the further difficulties in feeding living animals after they had been transported, others settled for dead ones, stuffed or skeletal. But the desire to collect the animals as well as the plants does come out in the work of artists, who were not constrained by the necessity to practise what they were, so to speak, recommending, and nowhere is this identification of the existing Botanic Garden with the original Edenic ideal shown to more striking effect than in one of the many illustrations to G. B. Andreini's *L'Adamo, sacra rapresentatione* published in 1617. Like du Bartas and Milton, Andreini was concerned with the themes of paradise lost and paradise regained, and at the very beginning of the book the Garden of Eden is shown as a typical enclosed garden of the period, with the many small, well trimmed beds described by Spenser, and a variety of wild animals sitting and strolling just outside the gate (Plate 39). In much the same way among the illustrations to Francis Quarles's emblems we find animals disporting themselves in a perfectly ordinary parterre, divided into four quarters, of a kind which must have been familiar to all his readers.[61]

The idea behind the Botanic Gardens and the more intelligently conceived among the great gardens of the day, was the recovery of knowledge, and of power over nature. By bringing all the plants and animals into one place one could name them, and by naming them men could communicate the nutritive and medicinal properties of the plants to each other, and render the animals both docile and serviceable. In the middle ages opinions differed much as to the amount of wisdom retained by man after the Fall and the capacity of men to recover what had been lost. But it was generally accepted that Solomon had possessed Adam's powers, John Cassian and Isidore of Seville seem to have believed that men could still raise themselves by their own efforts, and Roger of Wendover believed in the gradual rehabilitation of the human race through the just deeds of Adam's descendants.[62] Now, in the seventeenth century, du Bartas thought that God had left men with sufficient intelligence to subdue 'The stubborn'st heads of all the savage troupe',[63] while Parkinson thought that the capacity to comprehend natural truths had descended from Adam to Noah, and from Noah to his posterity,[64] and Evelyn said that 'Adam instructed his posteritie how to handle the spade so dextrously, that, in processe of tyme, men began, with the indulgence of heaven, to recover that, by Arte and Industrie, which was before produced to them spontaneously.'[65] Man was on the road back, and no one shared in this belief more fully than Francis Bacon, who was bent on the attainment of the Biblical promise to man of dominion over nature – an idea enshrined in the concept of Solomon's House in *The New Atlantis*. The whole great topic forms the subject of Charles Webster's book *The Great Instauration*, and Webster himself noted that 'Perhaps more than any other subject horticulture illustrated for Bacon the potentialities of the search for magnalia in nature.'[66] In the days before the industrial revolution the road back lay through a garden.

Reading in the book of God's works, the value of a Botanic Garden was that it

39. Animals before the entrance to the Garden of Eden.
   G. B. Andreini *L'Adamo, sacra rapresentatione*, 1617.

conveyed a direct knowledge of God. Since each plant was a created thing, and God had revealed a part of himself in each thing that he created, a complete collection of all the things created by God must reveal God completely. Given the supposed relation between the macrocosm and the microcosm, the man who knew nature best knew most about himself. The anatomy of man was the anatomy of the whole world, and as Stephen Blake argued, the man who 'could rightly know himself . . . might comprehend all things else in the Creation, yea the Creator himself'.[67] It did not matter, then, where one started. God, man, and the works of God in nature were all related, and the man who could unravel the secrets of one would understand the others. What was new at the time was the feeling that, the attempt to approach God by direct assault through the study of theology having failed, the place to start the inquiry was among the plants in a Botanic Garden. That being so, there was one very important consideration, which must surely have been seized on at the time, and that was that

> In nature's book the weakest brain may speed,
> Th' untaught may learn it, and th' unlettered read.[68]

The implications of this were very far-reaching, though they did not all tell one way. For the reactionary the implication was that there was no call to teach the poor and the illiterate to read. They could listen to the first of God's two books in

55

Church on Sundays, and they could 'read' in the other upon each of the remaining six days of the week throughout their lives. To the revolutionary the implication was that since the poor were as well able to read in the book of God's works as the rich, and might even have more opportunity to do so, their judgement was to be trusted. But here one is bound to conclude that the advantage lay, or should have lain, with the academic. This was not because, as Parkinson put it, the delight in the wonders of nature 'hath ever beene powerfull over dull, unnurtured, rusticke and savage people', and ought therefore to be more powerful still 'in the mindes of generous persons',[69] but because, from the very nature of their lives, hemmed in by laws of settlement, the experience of the poor was but partial, and was likely to be limited to their own locality and to a single page or two of nature's book. It was the man who could read most, who had most pages open to him, who could learn most. As Sir Thomas Browne said, 'the finger of God hath left an inscription upon all his works', and these works and the inscriptions upon them, when 'aptly joined together, do make one word that doth express their natures.'[70] That was why Cortusius standing in the garden at Padua, and naming the plants, was like a new Adam,[71] and that was why the unidentified R.I. praised Stephens and Brown, the authors of the catalogue of plants in the Botanic Garden at Oxford. Their work in naming the plants reminded him of the time when Adam was gardener to God's own paradise.

> He knew their names (for he their names did give)
> Of Fish and Foule, and what on earth did live.
> All Plants to *Adam* then, as now,
> *Flos Solis* to the Sun, their heads did bow.
> He was their Lord and could command a view
> As of their faces, so their natures too.[72]

Or, as 'Vertumnus' put it in his *Epistle to Mr Jacob Bobart*, the gardener at Oxford

> From Thee, the *Gods* no Knowledge hide,
> No Knowledge have to Thee deny'd.[73]

CHAPTER V

PHYSICK

THE Botanic Garden which was to serve the great general purpose of rediscovering the many faces of God in the creation, also fulfilled the more specific purpose of a physick garden or a collection of 'simples' (Plate 40). Ezekiel had spoken of the 'leaf' being for medicine.[1] All plants were believed to contain 'virtues' or healing powers, and in the garden into which plants had been gathered from all over the world there would be no hurt without a heal. At Padua the curator in charge of the Botanic Garden was the professor of pharmacology,[2] and the pillars supporting the tank from which the garden was watered featured statues of Aesculapius, Hippocrates, and Galen, three giants of ancient medicine, and of Mithridates, who was immune to poison – a useful characteristic in the Italy of the Borgias (Plates 41–2). At Amsterdam the two professors lectured upon the medicinal properties (Plate 43), one of the exotic, the other of the native plants,[3] while at Paris, the *Jardin du Roi*, which was also known as the *Jardin Royal des Plantes Médecinales*, was under the superintendence of the king's physician, and three doctors were hired to instruct all comers in the healing properties of the plants.[4] In the late sixteenth century when Gerard sought to interest Burghley in the foundation of a Botanic Garden at Cambridge he used the argument that it would help to make the 'noble science of physicke . . . absolute',[5] and early in the seventeenth century the Earl of Danby was motivated to establish the Botanic Garden at Oxford by the hope of promoting learning, 'especially the faculty of medicine'.[6] R.I. later said that the catalogue of the plants in the Botanic Garden at Oxford was 'one great Panax', and included 'Herbs fit to heale all sores, and cure all paines'.[7] Evelyn, who thought that 'without some tincture in Medicine,' gardening was a voluptuous and empty speculation, spoke of his Botanic or Philosophical garden, where he hoped to 'enfranchize' strange plants from all over the world and acclimatise them to the English air, as a medical garden, whence he would obtain all the remedies which he needed for his family and for the neighbourhood.[8] Finally, the Chelsea Physick Garden, which was established in 1673, was from the beginning intimately associated with the Society of Apothecaries, and seems always to have been more exclusively a garden of simples, and less of an encyclopaedic garden than the Botanic Garden at Oxford.

In this way, too, then, the Botanic Gardens were to '*Restore/* To us, in part, what *Adam* knew before'.[9] Cortusius was said to be occupied at Padua in recovering the lost knowledge of the virtues contained in the plants,[10] R.I., who referred to Adam as 'the great Simpler', declared Stephens and Brown's catalogue of the plants in the garden at Oxford to be a work worthy of Solomon himself,

> For yee, in this your learned worke expresse
> The Balme of Gilead in an English dresse,[11]

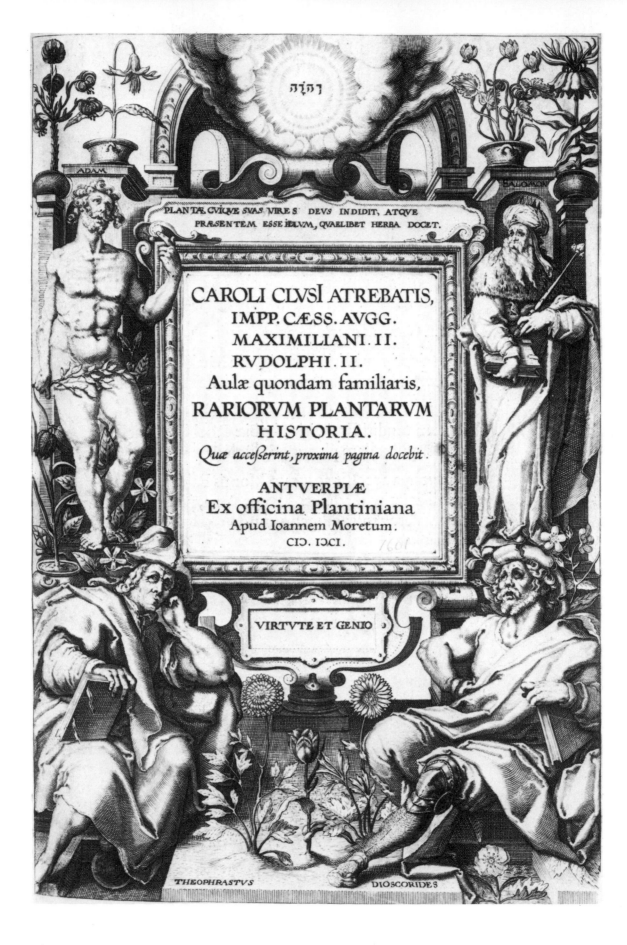

יהוה

ADAM

PLANTÆ CVIQVE SVAS VIRES DEVS INDIDIT, ATQVE
PRÆSENTEM ESSE IDLVM, QVÆLIBET HERBA DOCET.

CAROLI CLVSI ATREBATIS,
IMPP. CÆSS. AVGG.
MAXIMILIANI. II.
RVDOLPHI. II.
Aulæ quondam familiaris,
RARIORVM PLANTARVM
HISTORIA.
Quæ accesserint, proxima pagina docebit.

ANTVERPIÆ
Ex officina Plantiniana
Apud Ioannem Moretum.
CIƆ. IƆCI.

1601

VIRTVTE ET GENIO

THEOPHRASTVS   DIOSCORIDES

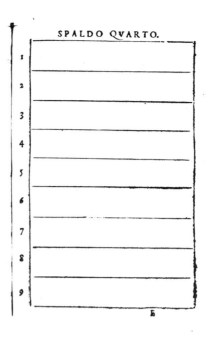

40. (facing page) The frontispiece to Clusius's *Rariorum plantarum historia*, 1601, brings together Adam, Solomon, Theophrastus the curator of Aristotle's Botanic Garden, and Dioscorides the classical authority upon pharmacology.

41. (above left) The fourth quarter of the Garden at Padua.

42. (above right) Blocks were printed with numbers corresponding to the beds in the Garden. Here the student could enter his notes.

    G. Porro *L'horto de i semplici di Padova*, 1591.

43. (below) The botany lesson in the Garden at Leyden (detail).

    P. Paaw *Hortus publicus academiae Lugdunum-Batavae*, 1601.

THEATRUM
BOTANICUM,
THE THEATER
OF PLANTES.
OR
An Universall and Compleate
HERBALL.

Composed by John Parkinson
Apothecarye of London, and the
Kings Herbarist.

LONDON.
Printed by Tho: Cotes.
1640.

ADAM.

SOLOMON.

W. Marshall sculpsit

and in his herbal *Adam in Eden; or, Nature's Paradise*, published in 1657, William Coles promised the reader to reinstate him 'into another *Eden*'.[12] The theology behind this use of language is never made explicit. R.I., who says that Adam lost his skill in herbs 'when he it needed most',[13] seems to imply that even in the original garden of Eden, Adam and Eve had suffered from the occasional head-ache, and had had recourse to healing plants. Jonstone on the other hand appears to have believed that God had disposed the medicinal plants outside the Garden, where they appeared, in due course when they were needed.[14] But this in turn should have raised the question, which he does not seem to have discussed, whether a far-seeing God, who had known that he was going to endow men with free will, had created medicinal plants because he knew men would fall. The vital questions whether there had been diseases in the Garden of Eden and whether these had been held in check by the practice of prelapsarian simpling, or whether all disease was a consequence of the Fall were left unanswered, as they were for example, by Joseph Fletcher, who, in his *Perfect-Cursed-Blessed man* hedged his bets and wrote of diseases 'or inveterate or fresh'.[15]

However that may be, belief in the medicinal properties of plants was universal. The labels beside the plants in the reconstructed early seventeenth-century garden at Kew, reveal that helleborus niger will get rid of melancholy, that symphytum tuberosum will relieve pains in the back from wrestling or overmuch use of women, and that gladiolus byzantinus will draw splinters out of the flesh. Du Bartas, who is probably as reliable a guide as any to the commonplace beliefs of his time, pointed out that other animals seemed to know by instinct where to find the remedies they required.

> Yet each of them can naturally finde
> What Simples cure the sicknesse of their kinde,
> Feeling no sooner their disease begin
> But they as soone have readie medicine.[16]

The ram ate rue when it felt unwell, and the tortoise hemlock, and man, too, should find his remedies among the plants, for

> God not content t' have given these Plants of ours
> Precious Perfumes, Fruites plenty, pleasant Flowers,
> Infused Phisike in their leaves and Mores
> To cure our sicknes, and to salve our sores.[17]

Du Bartas went on to recommend paeony for giddiness in infants, mugwort for sunstroke, sow-bread for women in labour, blue succory for bad eyesight, and aconite, which was itself a poison, for snakes' bites,

> O valiant Venome! O couragious Plant!
> Disdainfull Poyson! noble Combatant!

Referring to a different kind of problem, he advised betonie for breaking friend-ships off, and willow wort for making enemies shake hands.[18]

Colour Plate IV. John Parkinson's *Theatrum Botanicum* (1640) joins Adam to Solomon, who possessed Adam's knowledge, and both to the four continents, Europe with her horses and chariot, Asia with its rhinoceros, Africa with its zebra, and the newly discovered America with its llama, cacti, and passion flower. Copyright, Bodleian Library (B. 1. 17 Med.).

Abraham Cowley was also deeply attracted to the view that nature provided a remedy for every disease. Rocket, which was a powerful aid to married couples making love, should be avoided by 'chast Lads, and Girls', mint would act as a contraceptive, depriving the seed of all virtue, and turning brisk men into 'dull frigid Eunuchs', wormwood was effective against indigestion, sage remedied hangovers of all kinds, and baum acted as a tranquillizer, banishing all clouds and storms from the mind and leaving serenity and peace behind.[19] Cowley's most extravagant expectations, however, and those most relevant to this essay, were pinned upon the mistletoe, a remarkable plant, whose leaves were believed to draw their sustenance direct from the air and to be 'the Product of mere Nature'. This facility, which connected the plant with the prelapsarian world, suggested to Cowley that it, rather than du Bartas's aconite, disarmed serpents of their stings.[20]

The trouble with all this, apart from the technical problems of preparing the remedies and prescribing the dosages,[21] was that it was all much too general. Gerard and even Parkinson, who made a great point in his introduction to the *Theatrum Botanicum* in 1640, that with newly discovered plants coming in from the Indies almost every day, his herbal contained many species unknown to his predecessor, just listed the plants in alphabetical order, and if you wanted a remedy from the 1688 pages of the text of the *Theatrum Botanicum* you would not know where to begin – even with the aid of Parkinson's table of 'Vertues'. Dozens of plants were listed as being of use in the treatment of any particular disease, and many plants were described in such glowing terms that the reader might have been forgiven for supposing that they would cure anything. As John Hill said, 'the English herbals say so much we know not which to credit.'[22] In the middle ages some system had been introduced into the subject by associating plants, through the times at which they flowered, to the calendar of the Saints, and even now this line of inquiry was not quite dead.[23] But contemporaneously with the development of Botanic Gardens new attention was given to the ancient idea, that 'God hath imprinted upon the Plants, Herbs and Flowers, as it were in Hieroglyphics, the very signatures of their vertues'.[24] If God really had written upon the plants indications of the part of the body, or the nature of the disease they were intended to treat, then observation should lead to the conversion of medicine into an exact science.

The 'science' of 'signatures' had come down from the ancient world, and had been known, but not too highly regarded in the middle ages. But now, Gerard claimed to be acquainted with the 'properties and privie markes' of thousands of plants,[25] and William Coles thought to develop the discipline to the point where 'long obscured Wisdom' could be retrieved 'from her dark mists'.[26] Thus, to take one of the most widely credited examples, walnuts (Plate 44), according to Coles 'Have the perfect Signature of the Head: the outer husk or green Covering, represent the *Pericranium*, or outward skin of the skull, . . . and therefore salt[s] made of those husks or barks, are exceeding good for wounds in the head . . . The *Kernel* hath the very figure of the Brain, and therefore it is very profitable for the Brain . . .'[27] In a similar way either 'a decoction of the long Mosse that hangs upon Trees', or maidenhair fern, was good for the bald, a belief shared by Abraham Cowley, who, writing presumably for young savants, said

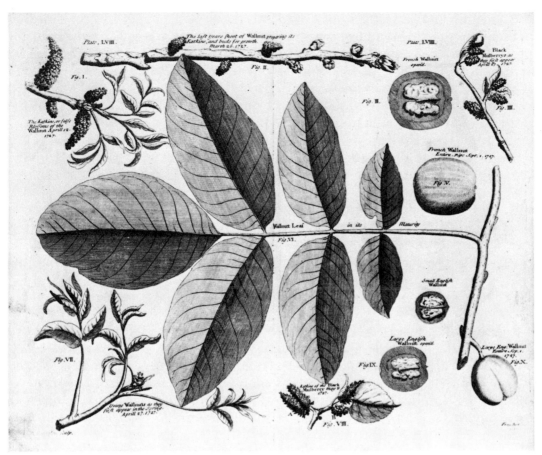

44. The walnut was believed to bear a 'signature'. Coles said that 'the *Kernel* hath the very figure of the Brain, and therefore it is very profitable for the Brain', but did not illustrate the tree. This plate comes from *Pomona*, 1729, by Batty Langley, who was not a signaturist.

> For when your Head is bald, or Hair grows thin,
> In vain you boast of Treasures lodg'd within,
> The Women won't believe you, nor will prize
> Such Wealth; all Lovers ought to please the Eyes.[28]

Thus it was certain, as Porta of Naples put it, that 'what part of Man they resemble that they are good for.'[29] A plant with yellow flowers, like the celandine, would cure jaundice,[30] convolvulus, because it twines, was good for the intestines, and lupins were good for the spleen.[31] A man's '*Triple Regions*' had their remedies at hand in the triple leaved plants, like the strawberry, if you thought the signature lay in the leaves, or in plants that grow double bulbs, if you thought the signature lay in the roots.[32] Plants did indeed provide a complete pharmacopoeia; 'each cure' as 'Anthracius Botanophilus' said, 'Is plainly legible in *Signature*';[33] and Coles' *Adam in Eden: or, Natures Paradise* (1657) included for the first time in England, 'A table of the Appropriations, shewing for what Part every *Plant* is chiefly medicinable throughout the whole Body of Man'. Here you could look up the remedies for drawing splinters out of the fingers and for ruptures, for shortness of breath in the lungs and for stones in the bladder, and, if you were a woman, for provoking and for stopping 'the termes' and 'the whites', for breeding milk and for contracting the breasts. Coles' book was said to have '*Methodiz'd* the *Art*', '*Smooth'd* the *Way*,/And *cut it shorter*, by whole *Shelv's* of *Books*'.[34]

Not everyone was carried away by the doctrine of signatures, and even at the height of it as a fashion in the early seventeenth century, de la Brosse denounced it.[35] Critics fastened first upon the confusions that arose when poisonous plants were brought into the picture, and, instead of indicating what plants to employ, the alleged signatures pointed out what plants to avoid. They might have added that with standards of book production as low, and copying as shameless as they were in the seventeenth century, the same engravings were apparently sometimes used to illustrate two different plants, one benign and the other poisonous.[36] A second line of criticism concerned the difficulty of deciding in what part of the plant the signature was written. Did it appear in the flower, the seed head, the leaves, the roots, the juice of the crushed stem, the smell, or, as happened when long-lived plants were recommended for longevity, in some other attribute altogether?

Much ingenuity was shown by supporters of the doctrine in answering these and other charges. Thus Coles himself, recognising that there were many plants with well-known medicinal properties which did not appear to have been signed 'concluded that a certain number were endowed with signatures, in order to set man on the right track . . . the remainder were purposely left blank, in order to encourage his skill and resource',[37] and Crausius, writing in 1697, devoted a whole section of a book on signatures to abuses of what he still insisted was a science.[38] Gradually, the claims made on behalf of the doctrine of signatures became too absurd to credit. Jonstone did nothing to advance it by stating that 'if Infants before they be twelve weeks old, be anointed with the juice of Wormwood on their hands and feet . . . neither heat nor cold will ever trouble them during their life.' He then retailed superstitions about a root that gave wonderful strength for Venus: 'they say that if a Man make water on it, he is presently provoked. If Virgins do but sit on them in the fields; or Urine upon them, the Hymen is presently broken, as if they had known a Man'.[39]

That the gardener in his Botanic Garden was expected to know

> What *Herbs*, our Bodies will sustain
> Secure from Sickness, and from Pain:
> What *Plants*, protect us from the Rage
> Of blighting Time, and blasting Age,[40]

is not in question. The problem is to know how far those who taught in the early Botanic Gardens subscribed to the doctrine of signatures or opposed it. Not surprisingly, the greatest English botanist of the second half of the seventeenth century, John Ray, did not believe in signatures, though he described his work as a natural scientist as 'thinking God's thoughts after Him',[41] and entitled his best known work *The Wisdom of God Manifested in the Works of the Creation*. But a number of volumes touching upon the topic have survived among the books collected by the Sherardian Professors of Botany at Oxford University, including two copies of the master protagonist William Coles' *Adam in Eden*. Coles said that the best hours of his life had been spent 'in the Fields and in Physick Gardens, more especially in that Famous One at *Oxford*, where I made it a great part of my study to be experienced in the laudable art of Simpling'. He claimed to have enjoyed the encouragement of the Principal of Hart Hall, and of other scholars and it really does look as though, all too often, in the atmosphere of heady optimism following the discovery of America (Plate 45) and the gathering together of the scattered pieces of the creation, this short cut to health appealed to charlatan and philosopher alike, and did as much to retard as to advance the growth of science.[42]

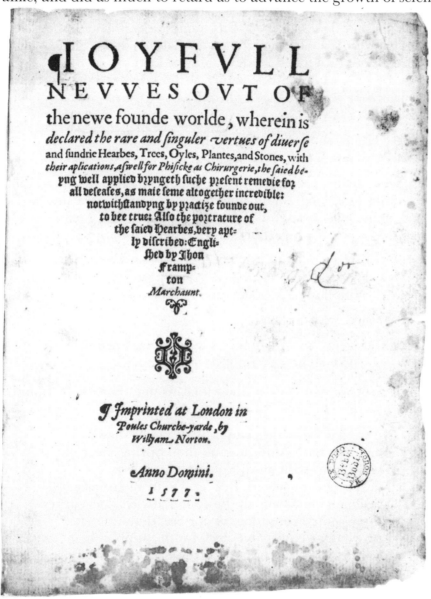

45. The title page of John Frampton's *Joyfull Newes*, 1577, reveals the hopes resting upon the healing properties of plants from the New World.

CHAPTER VI

SPRING AND FRUIT

IN their discussion of physick, writers of the early seventeenth century often leave it unclear whether man is compensating for his fallen condition as best he may, or whether he is, with raised expectations and higher hopes literally trying to reinstate himself into Paradise. Few such hesitations seem to inhibit contemporary writers in their treatment of the remaining aspects of the restoration of the original Garden of Eden and the innocence of its inhabitants, through the recovery of a perpetual Spring, the stress laid upon the use of fruit in diet, and the longing for an undivided human nature tuned to an indivisible God.

There can be no mistaking the longing of the late sixteenth and early seventeenth centuries for a perpetual Spring. In the *Faerie Queene* Spenser recounts how, in the Garden of Adonis

> They in that Garden planted be again,
> There is continual Spring, and Harvest there.[1]

Du Bartas, who, in his recitation of the Edenic theme, rehearses all the old questions, asking how long Adam and Eve spent in the Garden of Eden and whether they made love there, and (following Ovid) likening man's postlapsarian condition to that of a sailor at sea, imagines Eden as a happy seat with

> A climate temp'rate both for cold and heat,
> Which dainty *Flora* paveth sumptuouslie
> With flowrie Ver's innameld tapistrie.[2]

Marvell, in *Bermudas*, says that God

> gave us this eternal Spring,
> Which here enamells every thing.[3]

In his *Plantarum*, Abraham Cowley describes his new Fortunate Island in these terms:

> No Cold invades the temp'rate Summer there,
> More rich than Autumn, and than Spring more fair.
> The Months without distinction pass away,
> The Trees at once with Leaves, Fruit, Blossoms gay;
> The changing Moon all these, and always does survey.
> Nature some Fruits, does to our Soil deny,
> Nor what we have can ev'ry Month supply,
> But ev'ry Sort that happy Earth does bear,
> All Sorts it bears, and bears them all the Year.[4]

Milton, who confines his use of the term Paradise to the Garden of Eden, or earthly paradise, and calls its superior Heaven, sings the praise of 'vernal delight'. His

Garden of Eden is a place where 'Blossoms and fruits at once of golden hue/ Appeared',[5] and in Book V of *Paradise Lost*, when Eve is about to entertain Raphael to supper, we learn that fruit of 'All seasons, ripe for use hangs on the stalke',[6] so that when the table is laid there appears upon it

> All autumn piled, though spring and autumn here
> Danced hand in hand.[7]

The minor poet Habington writes that at one

> And the same season, Nature forth shall bring
> Riches of Autumne, pleasures of the Spring,[8]

and Hare, a diluvian and lapsarian controversialist, describes a Garden of Eden blest with 'a rich autumnal spring'.[9] Burnet himself, unusually lost for words, resorts to quoting a translation of Alcinus Avitus

> No change of Seasons or excess was there,
> No Winter chill'd, nor Summer scorch'd the Air,
> But with a constant Spring Nature was fresh and fair.[10]

These were the hopes, then, of poet and publicist alike, that the gardeners of the seventeenth century expected to be able to meet, when, writing in the context of God Almighty having been the first to plant a garden, they said, as Francis Bacon and John Evelyn did, that they sought to re-create a *ver perpetuum*.[11]

The re-creation of a perpetual Spring and the fruitful autumn that went with it depended upon several factors. First came the provision of shelter. In the *hortus conclusus* and in the Botanic Garden, shelter was provided by walls, which retained the heat even of the winter sun. But great men wanted to extend further than this. Writing for the gentry, Evelyn described the way in which fruit trees, properly planted in orchards, sheltered a house from both winter wind and summer sun.[12] More important still were evergreens, for these by the shelter they provided, helped, much more than bare winter orchards, to create a temperate climate. Since 'verdure' (Evelyn's word makes the point) was perpetually clad with leaves

> it defends both our Gardens and the dwelling from the penetration of the winds, and extreamities of the weather. And verily an ingenious Gardiner may so invirone his enclosures and Avenues, with Verdures, that they shall seem to be placed in one of the Summer Ilands and to enjoy an Eternal Spring, when all the rest of the country is bare and naked.

To this end Evelyn would have a gentleman plant the environs of a dwelling 'both with the taller and lower sort of Ever-greenes even for some miles about'. His seat would then become so infinitely delightful that 'the Winter Spectator' might imagine himself transported to one of the legendary islands in the Arabian Sea or in the West Indies, into Ethiopia, or to one of the Fortunate Islands.[13]

But the value of evergreens did not end with shelter. Green itself was associated with Spring and with life. Sir Philip Sidney's Arcadia was decked out in 'continuall greenes', Shakespeare linked the English legend of the green forest to that of the golden age in *As You Like It*,[14] and Hawkins referred to the Virgin as an 'evergreen Olive'.[15] Evergreens were simultaneously signs of the former, pre-lapsarian, uncorrupted nature (like parrots and the mistletoe), and symbols of eternal life, representing both the earthly and the heavenly paradises. That was a reason why they were grown in churchyards, and brought indoors at Christmas.[16] The imagery was strong and enduring. 'Vertumnus' rejoiced

> That *Trees*, in spite of Winds are seen,
> Array'd in Everlasting Green.[17]

and as late as 1770 William Hanbury, in *A Complete Body of Planting and Gardening*, recommended making two plantations, one, deciduous, in order to remind us of our end, and the other, evergreen, to impress upon us our immortal destiny.[18]

This second line of approach to the re-creation of a perpetual Spring by the provision of greenery, or foliage, could be carried a step further by making a collection of plants in such a way as to secure a regular succession of flowering. In Parkinson's day the most important means to hand to extend the flowering season into what had hitherto been regarded as the winter months, lay in the use of bulbs. Many of these

> shew forth their beauty and colours so early in the yeare, that they seeme to make a Garden of delight even in the Winter time, and doe so give their flowers one after another, that all their bravery is not fully spent, until that Gilliflowers, the pride of our English Gardens, do shew themselves: So that whosoever would have of every sort of these flowers, may have for every moneth severall colours and varieties, even from Christmas untill Midsommer, or after; and then, after some little respite, untill Christmas againe . . .[19]

Parkinson was no doubt exaggerating when he suggested that the most noticeable gap in the whole year would then come at midsummer. But this line of development was to be carried forward both far and rapidly. In 1713 'Vertumnus' interpreted the younger Bobart's task in the Botanic Garden at Oxford as being to ensure

> That *Flow'rs*, in spite of Frost and Snow,
> Thro'out our Year, perpetual Blow,[20]

and by 1746, James Harvey, in his *Reflections on a Flower Garden*, could suggest plants which he said would provide 'a sort of *immortal Corps* whose successionary Attendance never fails'.[21]

It cannot be said that either evergreens or bulbs were entirely new. Medieval England possessed the holly and the yew, box, and juniper, and upon the dissolution of the monasteries, Henry VIII's commissioners asked the gardeners at Hampton Court to take away the famous evergreens from the London Charterhouse – the cypresses, bays, and yews.[22] The Scots pine is indigenous, and among the bulbs, so, too, are certain species of snowdrop and daffodil. The monastic orders of medieval Europe were the multi-national corporations of their day: monks and nuns were great importers of new plants,[23] and it would not have been beyond the resources of the later medieval gardener, then, to have re-created some sort of perpetual Spring had he wanted to. But what seems to have happened

Colour Plate V. *Still Life* by Osias Beert the elder (*c.* 1580–1624), a painting which includes flowers from early Spring, the snowdrop, to mid-Summer, the rose, with a preference for the bulbs which were already turning gardens into scenes of continuous flowering. Copyright, Ashmolean Museum.

is that, with the discovery of America, and the introduction of completely new species, men suddenly became aware of possibilities which they had never really taken in before. Somewhere in the world were plants which flowered at every season of the year, however inclement the weather, and whatever theological improprieties they committed by doing so. The discovery of the new world stimulated a renewed search for hitherto unknown plants in Asia, where most of our modern bulbs come from, and even in Europe itself. The result was that the seventeenth-century gardener had a much greater range of materials to choose from. Evelyn's list of taller evergreens included abies, alaternus, arbor vitae, arbutus, celastrus, cedar, cypress, ilex, laurus, olive and picea, and his under-woods and thickets were to be planted with laurustinus, ligustrum, myrtus, oleander, phillyrea, pyracanthus, rosmarinus and ruscus, while Parkinson's pre-ferred 'outlandish' bulbs included the anemone, crocus, hyacinth and tulip, all of which have since been developed and marketed so successfully as to make every back-yard, in this sense, a potential Garden of Eden.[24] Finally, it is a striking fact that, in the seventeenth century, while gardeners laboured to re-create a perpetual Spring, Dutch and Flemish artists strove to achieve the same effect with still-life paintings in which spring and summer flowers, or spring flowers and autumn fruits appeared side by side (Col. Pl. V).

The restoration of a perpetual Spring was a prelude to the re-creation of man himself. The seventeenth-century gardeners were nothing if not optimistic, and there seems to have been no limit to the possibilities they now saw open before them. It was a commonplace that gardening was an 'innocent' occupation. Evelyn was but one among many who referred to it as the purest of human pleasures. Marvell shared his feelings, and in the mid-eighteenth century, Joseph Heely, extolling the virtues of Hagley, said that Littleton's garden was so beautiful that a 'villain would be disarmed from executing his dark and bloody purposes'.[25] Heely may have meant little more than a conventional eulogy, but had his remark been made a hundred years earlier, it would have been intended, quite literally, to convey the thought that gardens were capable of overcoming original sin.

When they wrote of innocence, many seventeenth-century writers really were trying to recover man's original, prelapsarian condition, and nowhere is this more evident than in the stress laid upon fruit as the original food of mankind, and upon the grafting of trees, the planting and care of orchards, and the transport and storage of fruit. The words garden and orchard were frequently treated as though they were synonymous, and Raleigh, Parkinson, Plat and Lawson, for example, all used some such phrase as 'Paradise is a garden or orchard'.[26] The orchard, like the garden, was surrounded by religious allegory. Solomon selected the apple tree as a figure of speech for his lovers,[27] Christian writers spoke of the church as God's garden or orchard, and artists portraying the martyrdom of St Stephen, who was stoned, showed the ground round about him strewn with apples to represent the fruitfulness of his sacrifice.[28] The seven virtues were thought of as trees planted in an orchard which was the Eden of the soul.[29] Cowley professed to see Eden in the autumn colours of a million orchards,[30] and Ralph Austen was only one of many who supposed that trees could talk, 'though not with *an articulate distinct voice*', yet 'very *intelligibly, and convincingly*', and in any language, '*to the minds, and consciences of*

46. 'The bird of Jove' and 'the beast that reigns in woods', from the first illustrated edition of *Paradise Lost*, 1688.

men'. Since the trees 'alwaies speake *the very truth*',[31] Austen chose to open his heart to the world in *The Spiritual Use of an Orchard* first published in 1653, and in *A Dialogue between the Husbandman and Fruit Trees* which appeared twenty-three years later.

It was generally agreed that, as Austen said, fruit 'was the first *Food* given to man',[32] and that until the Fall man ate fruit, while the beasts ate the herbs of the field, though du Bartas added honey and manna to man's original diet.[33] After the Fall the animals began to prey upon each other, and Milton interpreted this as the first sign Adam and Eve received as to the nature of their punishment (Plate 46).

> The bird of *Jove*, stooped from his airy tower,
> Two birds of gayest plume before him drove:
> Down from a hill the beast that reigns in woods,
> First hunter then, pursued a gentle brace,
> Goodliest of all the forest, hart and hinde.[34]

Colour Plate VI. (following page) 'Tradescant's Orchard' (*c.* 1611): the May cherry, which rivalled the early strawberry as the first fruit of the year. Copyright, Bodleian Library (Ashmole MS. 1461, plate 15).

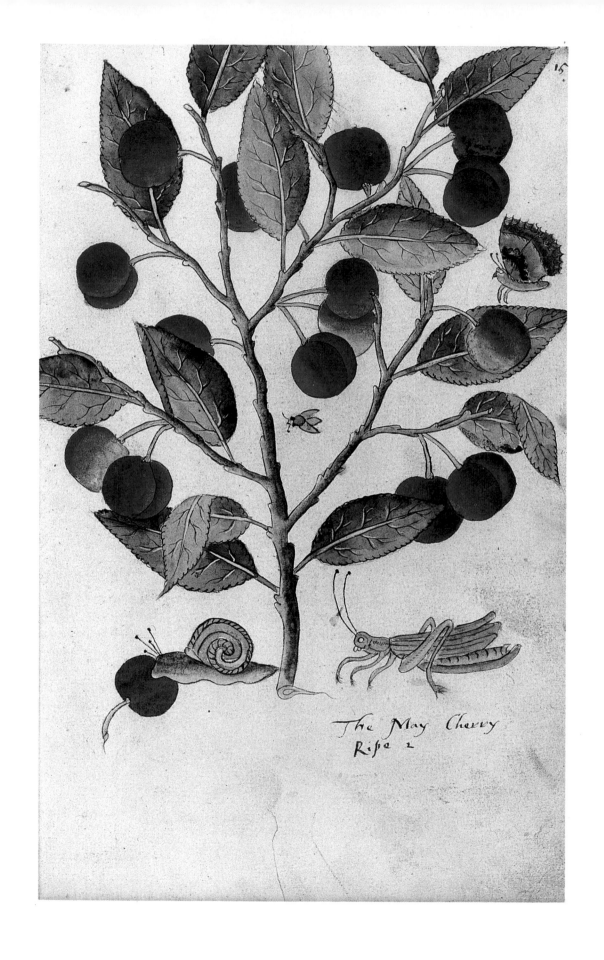

The May Cherry
Ripe 2

The peare Quince
Ripe October . 5 .

Milton, with his rather coy way of referring to the eagle and the lion, was attracted to the idea that man had added herbs to his diet between the Expulsion and the Flood, and had only become carnivorous after he came out of the Ark. This accounted for the decline from the first man and the first woman who were to live for ever, through the patriarchs, who lived for hundreds of years, to the ordinary mortals with their three score years and ten. The original fruit-eaters, who

> Nor any Food or Covering sought,
> But what from *Trees* and Woods they got,[35]

and whose work-load extended no further than just trimming or pruning the plants, ingested no decay. When men began to earn their living by the sweat of their brows, they had to supplement their diet with cereals which caused them to age, and finally, when they had to labour to restore a devastated world, meat became an essential part of their diet however short their lives became in consequence.

Throughout the middle ages monks were inclined to favour a meatless (but not a fishless) diet, and in the sixteenth and seventeenth centuries there were men who condemned the eating of meat, and became attracted to vegetarianism. Indeed, the wonder is that there were not more. In Mexico, the Franciscan, Mendieta, noted how the standard native diet of tortillas, herbs and chili 'did not create those "superfluous humours" which had to be ejected from the body in the form of anger, lust, greed, quarrelsomeness.'[36] Alas for him, in Mexico things were moving in the wrong direction: the natives were beginning to adopt European habits and to eat meat, and were becoming as wicked as the invaders. In England, in the seventeenth century, the vegetarian running seems to have been taken up by Ranters like John Robins, who prescribed vegetables and water for his disciples, by Roger Crab, the mad hatter of Chesham,[37] and by Thomas Tryon, who in his *Treatise of Cleanness* (1682) and his *Way of Health* (1683) referred to the animals as our fellow creatures, condemned the eating of flesh, and recommended 'the Lovers of Wisdom and Health to the more innocent use of *Grains, Fruits* and *Herbs*'.[38] But there is something not quite satisfactory about any of these three. Robins's disciples died of malnutrition; one of the motives of the mad hatter, who ended up on his hands and knees eating grass, was that of self-mortification; and Tryon was a compromiser, who, if his countrymen wouldn't give up meat, was quite willing to assist them by giving 'a particular Account of each sort of Flesh', in order that they might choose that which was least prejudicial to their health.[39] Tryon's vegetarianism, therefore, runs out in a commentary on the unhygienic conditions to be found in butchers' shops.

Tryon's experience was that, whatever they professed to believe, people simply would not give up meat. Although Shenstone still spoke, in the middle of the eighteenth century, of butcher's meat inflaming and stimulating the animal passions,[40] country gentlemen, and poets singing the life of a country gentleman, tended to approach the topic somewhat wistfully. Carew celebrated a time when

> The willing Oxe, of himselfe came
> Home to the slaughter, with the Lambe,
> And every beast did thither bring
> Himselfe to be an offering,[41]

Colour Plate VII. (preceding page) 'Tradescant's Orchard' (*c.* 1611): the peare quince, which brought the season's fruit to an end in October. Bodleian Library (Ashmole MS. 1461, plate 137) and Cowley went even further. Writing about a man in happy retirement from politics, the court, and a life of ambition and law-suits, he reversed the thrust of the traditional conception when he spoke of all his malice, all his craft, being shown 'In innocent Wars, on Beasts and Birds alone'.[42] The sentiment reminds one of the Schlaraffendland of Pieter Brueghel, where little pigs, already roasted – with knives and forks ready positioned in their flanks – run about squeaking 'eat me! eat me!'[43]

On the whole then, the animals did not benefit much from seventeenth-century concepts of an innocent diet. The appetite was too strong to be denied, and what happened was not that men abandoned meat, but that they consciously added fruit to their foodstuffs (Plate 47). The elder Bobart grew fruit-trees up the walls of the Botanic Garden in Oxford,[44] and the gardening literature of the period is filled with fruit, all the way from *The Fruiterer's Secret* of 1604, and *The Husbandman's Fruitful Orchard* of 1608, through Samuel Hartlib's *A Designe for Plentie, by an Universall Planting of Fruit Trees* (1652) and John Beale's *Treatise on Fruit Trees* (1653), to Evelyn's *Pomona*, and Sir William Temple's somewhat smug remark that 'all men eat fruit that can get it'.[45] Many other books which do not refer to fruit in the title, do, in fact, give pride of place to fruit in their contents, like Evelyn's *French Gardiner* (1658) and Leonard Meager's *English Gardner* (1670), and this concern with fruit was carried on into the eighteenth century with John Lawrence's *Clergyman's Recreation* (1714) and *Gentleman's Recreation* (1716) (Plate 48), and Samuel Collins' poem *Paradise Retriev'd* (1717), all of which were about the most beneficial methods of laying out an orchard, and managing and improving fruit trees. This interest in fruit trees spilled over into another whole branch of literature in praise of cider. Beer required hops (an imported plant that does not seem to have been universally welcomed): hops consumed timber, and succeeded, Evelyn asserted, but one year in three.[46] Evelyn, who thought that the smell of the blossom, even, of an orchard contributed to longevity (another prelapsarian idea), considered cider 'one of the most delicious and wholesome beverages in the whole world', and without rival as the one '*most eminent, soberly to exhilerate the* Spirits *of us* Hypocondriacal Islanders.'[47]

It is impossible to credit Evelyn's statement that the art of growing fruit-trees underwent a transformation in the reign of Henry VIII.[48] Fruit trees had been grafted for centuries, and what seems to have happened is not that an entirely new technique was introduced, but that in the orchard as in other areas of gardening, improved species were introduced from abroad. In 1611 when John Tradescant went to Flanders and France to buy plants for the Cecils' new garden at Hatfield, he travelled in September when the fruit hung on the trees, and the plants he bought – 450 from Holland and Belgium and nearly 500 from France, could be lifted and shipped immediately.[49] The catalogue of the fruit trees at Hatfield was known subsequently as 'Tradescant's Orchard', and contains sixty-four exquisite drawings of fruits including ten cherries, thirteen peaches and nineteen plums (Col. Pl. VI and VII).[50] All these new and improved strains of fruit imported from the continent would help to ensure better crops, and to narrow the gap between the last fruits of one year and the first of the next – between the last quinces and grapes and the first strawberries and cherries. Richard Bradley

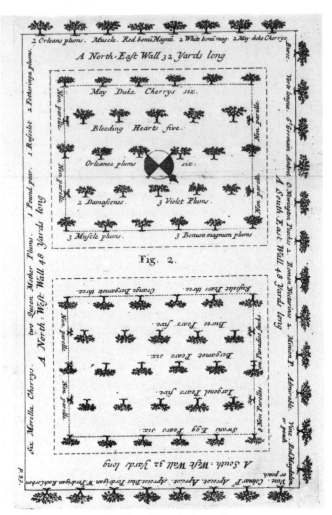

47. Citrus trees set out for the summer in the Garden at Leyden.
H. Boerhaave Index plantarum, 1710.

48. John Lawrence's plan for a fruit garden.
The Gentleman's Recreation, 1716.

boasted that in this way we could fill our gardens with fruit at every season of the year, but in practice there would still be a gap, to be filled either by storage or by the use of hothouses. Not surprisingly, then, *The Husbandman's Fruitful Orchard* gave advice on laying fruit up 'as may be for their best lasting and continuance', Gervase Markham's *English Husbandman* (1635) included a section on the preservation of all kinds of fruits, and Sir William Temple boasted that with storage, his estate at Sheen 'would furnish every day of the circling Year'.[51] Evelyn gave advice on drying fruit, on pickling it in salt and vinegar, and on preserving it in wine, cider and honey, and also struck a less than ideal note when, in writing about the conservation of fruits 'in their naturall', he suggested his readers should shelve the better sorts of fruit for themselves, leaving space on the floor in the attic for heaps of those which were commonest for the servants.[52] None of these authors, perhaps, realised that, even in the middle ages, apples had been stored right through the winter until the new season of soft fruit began again.[53]

With a further rush of blood to the head, the seventeenth-century gardeners seem next to have taken it for granted that any climate, however cold, could be

49. Flowers and fruits grow together on the evergreen orange tree, which raised it to Edenic status.
F. van Sterbeeck *Citricultura*, 1712.

50. The lemon possessed a somewhat ambiguous status as Adam's apple.
F. van Sterbeeck *Citricultura*, 1712.

made to blossom by the application of warmth. Centuries earlier Albertus Magnus claimed to have ripened fruit in January by the use of heat:[54] now James I imported mulberry trees in order to establish a silk industry, Leonard Meager claimed to be able to grow a salad fit to be brought to the table in three to four hours by putting it over a fire,[55] and rich men up and down the country began to take an interest not just in the erection of walls to store the heat of the sun, but in the provision of grates, flues and fires which would ward off even the sharpest frost. Many went much further still, and provided themselves with purpose-built houses for exotic plants and fruits – oranges and lemons especially (Plates 49, 50). These, as Cowley said

> were the fam'd *Hesperian* Fruits of old,
> Both Plants alike, ripe fruit and Blossoms hold.[56]

In the Netherlands the products of these orangeries may well have been a serious business enterprise, furnishing Dutch seamen with the precious juices that ward off scurvy.[57] In backward England, where casualties among seamen were held of little account, they remained the preserve of the rich. Sir William Temple boasted

I.

Totus mundus in maligno (mali ligno) positus est.
Will Marshall *sculp:*

51. (right) A typical temptation scene with the palm tree on the left presumably intended to represent the tree of life.
F. Quarles *Emblemes*, 1635.

52. (facing page) The palm tree was the favourite candidate for the tree of life. J. Pomet illustrated three kinds in *Le marchand sincère*, 1695.

that his orange trees were as large as any he saw in France, and the gardeners at Beddington claimed to have picked 10,000 oranges in a year from trees thirteen feet tall in a building two hundred feet long.[58] Stopping at nothing, gardeners even sought to grow pineapples, though it is far from clear when this was first achieved. Mea Allan thinks John Tradescant grew them at Oatlands for Charles I. Others suppose that the first pineapple grown in England was the one presented to Charles II, while Hugh Honour suggests that the first pineapple was not grown in England until 1720.[59] Perhaps it hardly matters: to the seventeenth-century gardener, living with the pre-industrial revolution idea of re-creation rather than the later concept of improvement or reform, nothing seemed impossible. Throw enough effort into the cultivation of the soil, import good stock, provide shelter by means of walls and evergreens, burn enough coal, which was becoming plentiful, to turn winter into spring, and there was nothing that could not be done as well in England as in the South of France. 'By Industry and Cultivation' the garden, as Hervey said as late as 1754, could be made 'an image of Eden'.[60]

Within their orchards and re-created Gardens of Eden, the optimists might well debate the identification of the tree of life. Had all fruit trees been trees of life, and did the tree of life, as Cowley suggested, still proffer its gift 'in the inexhaustible sustenance furnished by fig and date, the olive and the vine'?[61] Or had the tree of life been one particular tree, whose descendants (if they had suffered the common fate of living things and been scattered) might yet be discovered somewhere upon the globe? For two centuries the explorers cast long and lingering glances over the palm tree (Plate 51). The date palm had been known for centuries, and bore a

78

Gros Cocos dans leurs enveloppe.

Palmier portant les Dattes.

Arbre Portant les Cocos.

Noix Vomique.

Palmier des fruicts du quelle on tire l'Huile de Palme.

Fruiet du Palmier de l'Isle de Ceilan.

Noix de Maldive.

Cosos de divers grosseurs hors de leurs enveloppe.

Fagara.

Cocos Longs.

Coque de Levant.

wonderful reputation for enduring injury without complaint.[62] But the coconut palm was even more remarkable, because there was nothing, it seemed, that could not be obtained from it (Plate 52). Raleigh said that 'the Earth yeeldeth no plant comparable to this', and that this tree, which flourished in the perpetual spring of the East and West Indies 'giveth unto man whatsoever his life beggeth at Natures hand.'[63] Du Bartas was impressed by the way in which the one tree yielded wine, oil, vinegar, butter and sugar, while threads from the bark could be spun and woven into cloth like flax, and even the more scientific John Ray quoted with approval from the account of the coconut palm given in the *Hortus Malabaricus*.[64] Cowley's praise extended beyond the fruits of the tree to the products manufactured from its timber.

> What senseless Miser by the Gods abhor'd,
> Would covet more than *Coccus* doth afford?
> House, Garments, Beds and Boards, ev'n while we dine,
> Supplies both Meat and Dish, both Cup and Wine,
> Oil, Honey, Milk, the Stomach to delight,
> And poignant Sauce to whet the Appetite.
> Nor is her Service to the Land confin'd:
> For Ships intire compos'd of her we find,
> Sails, Tackle, Timber, Cables, Ribs and Mast
> Wherewith the Vessel fitted up, at last
> With her own Ware is freighted . . .[65]

53. Nobody was sure, but the banana was commonly identified as the tree of the knowledge of good and evil.

F. van Sterbeeck *Citricultura*, 1712.

Ironically, this was one tree which not even the greatest optimist could expect to see grown in an English garden. Even this, however, did not apparently put an end to the sense of expectation, running right through the seventeenth century, that somebody might yet discover, somewhere, a short and easy route to the recovery of innocence and eternal life.

Strange as it must seem, much less attention was paid to the problem of identifying the forbidden tree of the Garden of Eden. But the favourite candidate for the tree of the knowledge of good and evil was the banana (Plate 53). In 1597, Gerard, in his *Herball*, reported that the Jews 'suppose it to be that tree, of whose fruit Adam did taste'. Gerard himself was inclined to dismiss this as a fable, but as late as 1736 Linnaeus still considered this hypothetical identification worth a mention in his *Musa Cliffortiana*.[66]

The Botanic Garden and the orchard also served to realise the sexual ideals of what might be called a severely Christian age. Most of us, if asked to comment upon the story of the Fall, would say that it took place when Eve ate the forbidden fruit. But there was another tradition in Christian thought, in which it was maintained that the damage was done when the woman was taken out of man and created, and that the division between the sexes was the very first example of scattering and corruption. Ambrose thought that the best state was that of Adam before Eve was created, and John Scotus, Gregory of Nyssa, and John of Damascus all taught that the division between the sexes was the consequence of sin.[67] Røstvig has called attention to the prevalence of this line of thought throughout the seventeenth century, when Norris's ideal condition was that of Adam before Eve was created, 'Like the first man in Paradise alone'.[68] This so-called Hermetic doctrine comes over strongly in Marvell, who wanted to convert all eroticism into a green thought in a green shade, and longed to see the creature he loved metamorphosed into a vegetable being.[69] The happy creature was that which contained or comprehended both male and female natures. God was such a person, and since the sexual reproduction of plants seems to have gone almost completely unnoticed, plants, too, were thought to share this characteristic.

Both the Tradescants and Stephens and Brown distinguished between the male 'firre' tree and the female, and in France de la Brosse explicitly raised the question of the sexuality of plants.[70] But, the fir tree apart, sexual reproduction seems to have been universally recognised only in the Palm tree, meaning, in this case, the date palm. This was referred to time and again, by Hawkins, who pointed out that 'the Female never brings forth fruits, but standing opposit by her Male', by Cowley, who drew attention to the way in which the roots of a male and female palm tree embraced underground, and by John Ray, who was aware that the phenomenon of sexual reproduction in palm trees had been known at least since the days of Theophrastus, the keeper of Aristotle's Botanic Garden.[71] This sexual reproduction of palm trees was assumed to be exceptional, one of the wonderful things of nature as Jonstone put it, and was easily assimilated to theology and smothered in imagery. Because the female never brought forth fruit unless her male were by her, the palms were credited with being faithful lovers. Hawkins piled on the allegory, likening them to turtle doves, to 'constant *Ulysses*, and chast *Penelope*', and above all by referring to them as 'the Hieroglyphick of Nuptials', representing the 'marriages' of Christ to his mother, and to his church. The fact, if it be a fact, that the male tree grows faster than the female, was interpreted as reminding us that Christ died and reached Heaven before the death and assumption of the Virgin Mary.[72]

Palm trees apart, then, sexuality, as Albertus Magnus had said in the thirteenth century, was unknown in the plant world, and a material union of two plants referred not to the reproduction of the species, but to mutual cooperation of the kind offered by the elm tree to the vine.[73] Even a man like Jonstone, who referred to the sexual mating of palms, was not led on to consider the possibility that, far from being an exception drawn by God to point a moral, palms were, in fact, exceptional only in doing conspicuously what the generality of plants did in their seasons. He continued, therefore, to suppose that the principles of male and female were mingled in plants, and that they possessed vegetative souls.[74] Without saying anything quite so crude, then, as that God was a vegetable, or that vegetables were Gods, the two shared the vital characteristic of an undivided nature that was at peace with itself. Whereas animals were observed to reproduce sexually, and were regarded as having participated in the fate, if not the sin, of man, and believed to have divided natures, plants, on the other hand, were mirror images of God. According to the ideal of Christian universalism, that man would know the creator well who built himself a Botanic and a Zoological Garden whither he could import all the plants and animals, and contemplate the thousands of varieties of self-expression assumed by God. This was now changed to read that that man would know the Creator better who retired to a Botanic Garden, excluded all the animals, and not only all the animals, but all the women too. Marvell wrote of

> that happy Garden-state
> While man there walk'd without a mate.
> Two Paradises 'twere in one,
> To live in Paradise alone.[75]

A similar thought comes across clearly in R.I.'s address to Stephens and Brown, the authors of the catalogue of plants in the Botanic Garden at Oxford

> When first the Globe of Earth fram'd by the hand
> Of him, that made and doth all things command,
> Did cloth'd appear in robes of youthfull green;
> Man then untouch'd with woman or her sinne,
> Naked like Truth, as Truth without disguise,
> Was Gardner made to Gods own Paradize.

These sentiments were peculiarly appropriate to the male preserve of Oxford University. Nor were they in any way countered by Abraham Cowley's flight of fancy in Book II of the *Plantarum*, where he made the plants in the Botanic Garden, in a conclave presided over by Mugwort (Plate 54), dispute which was the most useful of all the medical herbs to 'succour Womens Pains'.[76] Since there was no herb capable of ending their secession, and reuniting them with men in a single nature, women were in more need of physick than men. The council ended appropriately enough when Robert (i.e. Bobart the gardener) came out into the

54. John Locke's specimen of mugwort, *Artemisia vulgaris*. 'The red stalks having the signature of Womens Flowers', there is, Coles asserted, 'no Herb so generally received . . . for the curing of Womens diseases'. Copyright, Bodleian Library (MSS Locke c. 41 fol. 515 R).

garden early in the morning to pick some Sow-bread (Plate 55), because his wife was in labour, and he wanted, naturally, 'with greater Ease' to 'prove a Father'.[77]

Similar considerations applied in the orchard. The crucial texts here were *Genesis* I:11, which speaks of 'the fruit tree yielding fruit after his kind, whose seed is in itself', and *Genesis* I:29, in which God said that he had given man 'every tree, in the which is the fruit of a tree yielding seed'. These verses drew attention to a phenomenon which was completely misunderstood to mean that fruits appeared asexually. To the ordinary man it appeared to be a fact that trees produced fruits, and that fruits contained seeds which would grow into more trees and yield yet more fruit. The fruits grew without Venus's act, and fruit was an innocent diet, not just because it was taken without cruelty, for so if it came to that was milk, but because it shared in and communicated the ideal of an undivided nature.

To this it should be added that orchards were also asexually ideal because they were the home of bees and because they attracted birds. There is irony in the fact that bees, which perform such a vital function as intermediaries in the sex lives of fruit trees, should in this way have been accorded a place of honour in an asexual ideal world. But so it was, because the bee, by virtue of its supposed lack of procreative ability, also symbolised the state of innocence,[78] and it is surely no accident that so many of the seventeenth-century manuals on orchards also include sections upon bees. But how, then, did it come about that birds, which do have sex lives (though they are often discreet about them), came to be encouraged into this asexual paradise too? The answer lay in the facts that birds have two legs, in which they resemble man,[79] and that they fly, in which they resemble angels. In the middle ages this was interpreted to mean that birds had not participated in the original revolt against God, and that as a reward they had been permitted to live in the air, closer to the heavenly paradise than other animals and even than man himself. Thus, while farmers exterminated them for doing damage, and parish officers paid head money for 'vermin', idealists were inclined to welcome them even if they did help themselves to the fruit, and perhaps even because they helped themselves to the fruit.[80] Du Bartas's Eden contained a thousand sorts of birds, Lawson said that 'You need not want their Companie if you have ripe Cherries or Berries', Milton's Paradise was full of avian choirs, and Evelyn, who expected an orchard to harbour 'a constant Aviary of sweet Singers', wanted to construct an aviary in his Botanic Garden large enough to hold five hundred small birds, including linnets and yellow hammers, finches, larks, thrushes, blackbirds and robins.[81] Even as late as the beginning of the eighteenth century, Addison, who wrote one of the great hymns to creation, said that he valued his garden more 'for being full of blackbirds than cherries,' and that he very frankly gave them fruit for their songs.[82] He wrote in a tradition going back to the middle ages, when birds' songs were interpreted as a continual reminder of the songs sung by Isaiah and the prophets.[83]

Finally, it is worth noticing just how far the substitution of the select orchard for the all inclusive garden, and the more rigorous and ideological exclusion of animals, and even women, from the enclosed orchard, could carry one from the ideals of christian universalism found in the Botanic Garden. Ralph Austen's *Spiritual Use of an Orchard or Garden of Fruit Trees* (Plate 56) shows all the righteous-

55. John Locke's specimens of sow-bread, *cyclaminus*. A glance at the leaves, Coles thought, was enough to show that the plant belonged to the womb by signature, 'and therefore as Theophrastus affirmeth, the fresh Root put into a Cloth, and applyed for a little time, to the secret parts of a woman, that is in sore and long Travail in Child-birth, helpeth . . . to an easie and speedy delivery.' Copyright, Bodleian Library (MSS Locke, b. 7 fol. 339 R, detail).

ness, the withdrawal symptoms and the siege mentality of the sects. Austen was a Calvinist, and was at one time a Proctor at Oxford University. His book is based upon the supposition that the stocks and grafts of the trees had literally been separated at the Fall, and allegorically reunited at the Passion. When they were expelled from the Garden of Eden, men were left in a world of wild crabs. Sixteen hundred years ago 'God the great husbandman of his orchard the church, began to plant the vast waste grounds, the wilderness of the idolatrous nations, the Gentiles.'[84] But God 'upon his own free will and pleasure, without any foresight of faith, repentance, good works or any thing in us', had 'from all eternity made choice of what Spiritual Plants he pleased to plant in his Garden, the Church, and refused others.' God had 'left other plants in the woods and waste grounds', and Austen did not think we should try to convert them ourselves.[85] There are many wicked men and women in the world, many of them bear fruits which are beautiful to look on, though 'very harsh' and 'sower' to the taste,[86] and so it was he says, among us of late years: 'bowing at the name of Jesus, and communion table, surplesse, common prayer &c.'[87]

The more decisive the difference between the orchard of select trees and the wilderness the better. This is the gardening of the elect. The Saints within their enclosure will contemplate their fruit trees, and reflect upon the way in which 'in all persons regenerate, there are two natures, the one contrary to the other, the spirit and the flesh',[88] and the way in which the believer who is truly engrafted into Christ will yield good fruit.[89] But the stock of a fruit tree is the wild tree, and although sap and life come from the roots,[90] the graft is, as Austen himself notes, the superior part, joined onto the other to overrule it.[91] However well grounded, theologically, in a conflation of *John* XV and *Romans* XI, Austen's use of allegory thus runs into hopeless confusion. He himself recognises this,[92] but quite undeterred, continues to proclaim that behind their wall, the Saints have assurance of the pardon of sin, freedom from Hell, and Satan.[93] In their orchard they contemplate the signs of the times, and the imminence of the Apocalypse. When, he asks 'were the heavens, and the earth and the sea so shaken as they have been of late years?'[94] God had cut down flourishing trees like Charles I, and 'When God beginnes to reforme a nation . . . its a signe he intends to dwell there'.[95] So, just as Columbus believed that the gospel must be preached to the ends of the earth before Christ came a second time to restore the faithful to the earthly paradise, and that God had made him the messenger of the new heaven and the new earth,[96] Austen, secure within his orchard, sought to manicure the branches so that not a single spur would be left unpruned, or found out of place, when Christ came to commend his faithful servants and to 'delight himself in his garden.'[97]

56. Ralph Austen's *Treatise*, or *Spiritual Use*, 1657, looked to the new heaven and the new earth that were to follow the apocalypse rather than to the re-creation of the Garden of Eden.

CHAPTER VII

NATURE

O PTIMISTIC as the great gardeners of the seventeenth century were, they had to express themselves, and to seek to realise their ideals in a society that believed in original sin. Speculation as to the cause of the Fall, and the desire to catalogue its consequences still continued at a high level. Men's disobedience, and the reasons for it, were the great themes of the poets. Du Bartas laid great stress upon man's explicit acceptance of the prohibition, whose transgression would deserve 'sharpest Judge' rather than 'mildnes of a Father'.[1] Cowley attributed man's having 'Disdained Obedience to the Sovereign Guide' to pride,[2] and Milton's whole epic, of course, sang of man's first disobedience in ways too subtle to be summarised in a single sentence. When we turn to the consequences, du Bartas makes Adam wake up, crying 'This is not the World: O whither am I brought?' It is not '*The First-Weekes* glorious workmanship': the days are reducing, the nights increasing,

> his haile, his raine, his frost and heate,
> Doth partch, and pinch, and over-whelme, and beate.

The very earth has become barren and ungrateful

> Mocking our hopes, turning our seed-Wheat-kernel
> To burne-graine Thistle, and to vapourie Darnel.

Poisonous plants appear, the plants compete against each other in enmity, and the animals begin to tear each other to pieces.[3] These themes were given classic expression in Joseph Fletcher's history of *The Perfect-Cursed-Blessed Man* published in 1628. The penalty for disobedience was that God beset the man with woes

> Making all Natures Children turn his foes.
> 'Cause Man Himselfe from God was now declin'd,
> God made the Creatures all goe out of *kinde*.
> He curst the *Ground*, or with *Sterility*
> Or else with hurtfull weeds *fertility*.
> . . .
> The Living-creatures also, once all tame,
> Now refractory, and all wilde became.[4]

The conviction remained, then, that the whole world had been ruined at the Fall. As du Bartas put it, since his sin 'the woefull wretch findes none', stone, herb or beast, field, stream or mountain, 'But beares his Deaths-doombe openly ingraven'.

> In brief, the whole scope this round Center hath,
> Is a true store-house of Heav'ns righteous wrath.[5]

To Shakespeare the seasons' difference was a consequence of the Fall, and Marvell caught the general feeling perfectly when he wrote,

'Tis not, what once it was, the *World*
But a rude heap together hurl'd;
All negligently overthrown,
Gulfes, Deserts, Precipices, Stone,[6]

while Brome, being a little more specific topographically, described Derbyshire as

this durty corner of the World,
Where all the rubbish of the rest is hurl'd.[7]

Reynolds wrote of all nature having been poisoned by the Fall,[8] Harvey, in his adaptation of Benedict von Haefton's *Schola Cordis*, makes his speaker complain that the world is full of wild beasts,[9] George Fox, the Quaker, described the world as 'a briary, thorny wilderness',[10] and Evelyn referred to moss, rushes, wild tansy, sedge, flags, ferns and yarrow as auguries of a 'cursed' soil.[11]

Poet and theologian alike continued to rationalise this state of affairs through the doctrine of *felix culpa*. Just as Dante had described the Garden of Eden as a way-station on the road to a greater end,[12] so du Bartas advised man that

but that thou didst erre,
CHRIST had not come . . .
Making thee blessed more since thine offence,
Than in thy primer happy innocence.[13]

Richard Baxter, in *The Saints Everlasting Rest*, said that 'If Man had kept his first Rest in Paradise, God had not had opportunity to manifest that far greater love to the World in the giving of his Son',[14] and Milton, whose whole poem was written to justify the ways of God to man, concluded with a vision of the day when Christ would re-establish the earthly paradise,

far happier place
Than this of Eden, and far happier days.[15]

This preoccupation with the rationalisation of the Fall and its consequences, came down from the middle ages to the Renaissance as a living tradition, and in his emblem book Francis Quarles showed how a man sitting in the shade of a tree could keep his eyes fixed on the crucifixion (Plate 57), and a man sitting dallying with a naked woman and gazing down a short early seventeenth-century avenue should be able to see, at the far end, the Devil, with the flames of hell behind him, and smoke rising to darken the throne of Christ and his angels above (Plate 58).[16]

It was in an intellectual climate still filled with religious allegory that the existing form of the enclosed garden, or *hortus conclusus*, which had begun as a place of refuge and a scene of *contemptus mundi* (and still remained one to the saints in their orchards), was adopted by the sixteenth- and seventeenth-century gardeners when they came to construct Botanic Gardens. The most obvious feature of the layout of these gardens is that they were ordered. They were laid out at one time, they were regular in shape, they were level, many of them were filled with a good mixed soil, and they were planted in a regular manner. They were laid out in a single operation because it was obvious from the account in *Genesis* that God had made the Garden of Eden in a day, and if, in fact, it was not possible to make a

57. The Crucifixion seen in the imagination in the branches of a fruit tree.

    F. Quarles *Emblemes*, 1635.

58. Dalliance at one end of the avenue, and death (visible if the telescope were the right way round), at the other.

    F. Quarles *Emblemes*, 1635.

Botanic Garden in a day, or even in a year, then it should at least all be designed at one time. Thus the Botanic Garden at Padua was founded and apparently planned in 1545, and over forty years later Porro, reporting that it wasn't finished yet, added that they were following the 'blueprint' out, from the centre towards the circumference.[17] At Leyden and Oxford the work seems to have proceeded according to a draft, and as late as 1721 Richard Bradley regretted that the Director of the gardens at Kensington had not had the good fortune to lay the different areas out simultaneously.[18]

    The gardens were regular in shape because many people living in a disordered world believed that the world had been regularly laid out by God before the Fall. There were those who held that the 'orbicular or round forme' was the most absolute and perfect.[19] It was readily adaptable to either pre-Copernican or post-Copernican astronomy, and the round form was chosen for the garden at Padua. But the general preference was for the four square. Perhaps this was an example of men rationalising what they could not avoid, for as both Boyceau and Parkinson said, the practical advantages of the rectangle are considerable, and must be readily apparent to a surveyor, let alone a property developer.[20] In the second half of the seventeenth century, Evelyn, Sir William Temple, and John Worlidge all thought a square or rectangle the best,[21] and it is noticeable that the garden at

Leyden was rectangular, while that at Oxford was square, and even the circular walls of the garden at Padua contained a square. Ralph Austen, in his orchard, who disliked regular shapes because they reminded him of the Laws of Moses from which Christ had set us free,[22] was a lonely exception, and even he ended up with a scrupulously disciplined rectangular orchard for a frontispiece to his *Spiritual Use*.

The planned and regularly shaped garden should also be levelled and filled with a good mixed soil. When mountains were thought of as warts, and many people believed that the world, when first created, was perfectly level and smooth, it followed that a garden of re-creation must be levelled, and at Oxford '4000 loads of mucke and dunge' were brought in by the University Scavenger to raise and level the chosen site outside the town wall on the North bank of the Cherwell river.[23] It was a further consequence of the belief in a disordered world that men supposed the original fruitful soil to have been separated, into sands and clays, limes and peats, which must now be brought together again and mixed in order to restore the original fertility. Just as in agriculture one marled the fields in order to improve the crops, so, in a Botanic Garden, one must create a good rich loam. Richard Bradley expressed himself in favour of the improvement of one soil by another,[24] though de la Brosse followed a more sophisticated approach, keeping the soils apart because he aimed to reproduce as many environments as possible, and Evelyn, recognising that plants were now adapted to their environment, aimed to make up each step on the mount in his Elysium with a different soil.[25] Finally, this garden that had been laid out within a regular frame, levelled, and filled with a good mixed soil, should itself be planted in a regular manner. Du Bartas described a Garden of Eden laid out in the formal love-knots, triangles and lozenges favoured in the late sixteenth century,[26] and Cowley's medicinal herbs, when they held their conclave in the Botanical Garden at Oxford,

> met upon a Bed, neat smooth and round,
> And softly sate in Order on the Ground.[27]

George Herbert, in his poem *Paradise*, with its peculiar diminishing symmetry, line by line, wrote

> I Blesse thee, Lord, because I GROW
> Among thy trees, which in a   ROW
> To thee both fruit and order   ow,[28]

and this theme of order among the trees is common throughout the seventeenth century. Cowley wrote of groves 'rang'd all a row',[29] Evelyn toyed with the idea that the derivation of the word garden meant that everything in Paradise had been planted in straight lines,[30] Sir Thomas Browne wanted to believe that the trees in the Garden of Eden had been planted in quincunx,[31] and 'Vertumnus' held to the traditional theme when he stressed how, at Oxford, thanks to Danby's benefaction, trees, shrubs, and plants had been ranged in order, whereupon '*Eden* from the *Chaos* rose'.[32]

The Botanic Garden ideal, combining Christian universalism with the *hortus conclusus*, did not outlast the century. It is difficult to know how far to ascribe this to intellectual factors like the failure of the doctrine of signatures to establish itself as a science, to the shift in medicine to a 'chemical' (i.e. mineral) basis, and to Nehemiah Grew's discovery of the sexuality of plants. This last was taken up by Linnaeus, in the eighteenth century, and triumphantly employed to separate the plants into their various tribes and families – to classify them. Viewed from the twentieth century there is a supreme irony in Linnaeus's achievement. The seventeenth-century idealist had thought of plants as being untroubled by sexuality. Linnaeus's observations of their sexual organs enabled him to range the plants in order, the task which the founders of the Botanic Gardens had set themselves to achieve. But in his own time Linnaeus's achievements were accepted but slowly: God could not have made the plants promiscuous. The Linnaean system was still disputed by Tournefort in the middle of the eighteenth century, it was not adopted by Philip Miller of the Chelsea Physick Garden until 1759, and even then John Hill thought that there remained 'a truly natural method' of ordering plants, waiting to be discovered. This would be one 'in which the families established by nature should never be divided; her characters explained, and the gradations shewn, by which she passes from one class, and even one genus, to another'. This was 'the great desideratum' in the study of plants, and there was, therefore, as he argued, need for the construction of yet another new Botanic Garden in England to rival the *Jardin du Roi* in Paris.[33]

There were, however, two relatively matter of fact reasons for the decline of the ideal. In the first place, in the enthusiasm engendered by the discovery of the new world, the idealists had miscalculated the sheer number of plants that would be involved. We catch a glimpse of this in Cowley's *Plantarum* during the dispute between the continents, when the Indian God, who was upset because he

> heard them nothing say
> Of Fruits that grow in his *America*,

reproached the Europeans by pointing out that

> We still have many to our Conqu'rors Shame,
> Of which you are as yet to learn the Name.[34]

Before the century was out John Ray drew attention to the fact that Bauhin, the greatest of the Renaissance botanists, had catalogued over six thousand plants, and estimated that there must still be twice as many more waiting to be discovered.[35] After that, the days when one could describe the Botanic Garden at Oxford, with its two thousand different species in 1658, as the world in a chamber, were numbered.

In the second place, the gardeners of the early seventeenth century had been over optimistic, though they were scarcely to have known it. Climatic change was against them, and northern Europe headed back into a 'little ice age' in the second half of the seventeenth century.[36] Winters became more severe, and the expense of growing fruit in glasshouses became ever more prohibitive except for the very rich, so that, whatever the apple and the pear might do out of doors, the more exotic fruits like oranges and lemons, did not yield fruit equally, as Austen had said fruit trees should, to the rich and to the poor.[37] Furthermore, either because the climate was deteriorating, or because it never had been adequate, many of the plants imported from southern climes did not grow well, however much attention was lavished on them, and some did not grow at all.

Stay-at-home botanists had to be content to become acquainted with many of the more exotic plants from specimens preserved in an herbarium, or in what was known as a *hortus siccus*. This led to a remarkable shift of emphasis between the time of Parkinson and that of Ray. In the early seventeenth century the most esteemed gardener was the collector, who worked in his garden waiting for other people to bring plants to him: in the second half of the century the model gardener became the travelling botanist, the man who went to seek the plants out in their environment, where they grew best. The London Apothecaries, especially Thomas Johnson, whom Canon Raven described as the first man to explore England systematically for plants,[38] organised trips into the country to train their apprentices to identify the herbs used in their craft. John Ray's career began with expeditions into the fields round Cambridge, which were gradually extended to other counties until they became the *Catalogus plantarum Angliae* of 1670. This was followed by the *Observations . . . made in a journey through part of the Low-Countries, Germany, Italy and France: with a catalogue of plants not native of England, found spontaneously growing in those parts, and their virtues*, to which there was added an account by Francis Willoughby of a journey through Spain. Among the learned, the object now was to observe plants in their native habitats, and in the world of business and commerce there was no more talk of growing mulberries and silkworms, and establishing a silk industry in Britain. Plants like sugar, rice and tobacco were grown where they grew well, and not where it would be an expensive triumph to grow them at all. Division of labour in the growth of crops, and ocean transport, became the order of the day, not encyclopaedic independence and autarky.

But perhaps the most important factor leading to the decline of the Botanic Garden ideal was the fact that the optimism of its enthusiasts knew no bounds and could not be for ever contained within the narrow compass of the *hortus conclusus*. For those who did not share the restrictive mentality of Ralph Austen, there was no limit to the amount of land that could be recovered from the wilderness and from the consequences of the Fall. In Gabriel Plattes' imaginary land of Macaria, emphasis was placed upon the means by which 'the whole kingdom is become like to a fruitfull Garden'.[39] In 1652 Samuel Hartlib produced *A Designe for Plentie, by an Universall Planting of Fruit Trees*. Every headland, and every piece of waste ground was to be enriched with apples, pears, quinces and walnuts, 'for the relief of the poor, the benefit of the rich, and the delight of all', and the commons were to be planted with a tree every thirty yards.[40] John Evelyn copied this idea when he proposed to plant a fruit tree every one hundred feet throughout the country, and went further by proposing to surround London with regular enclosures of thirty to forty acres in extent filled with sweet-scented flowers.[41] The very title of Timothy Nourse's *Campania Felix*, published in 1700, where he wrote that gardening might 'most properly be call'd a Recreation . . . from the Restoration of Nature',[42] conveys the expansive outlook, and a few years later Charles Evelyn echoed this sentiment with his call to make the fields and meadows an open garden 'and the whole Country a perfect Paradise'.[43]

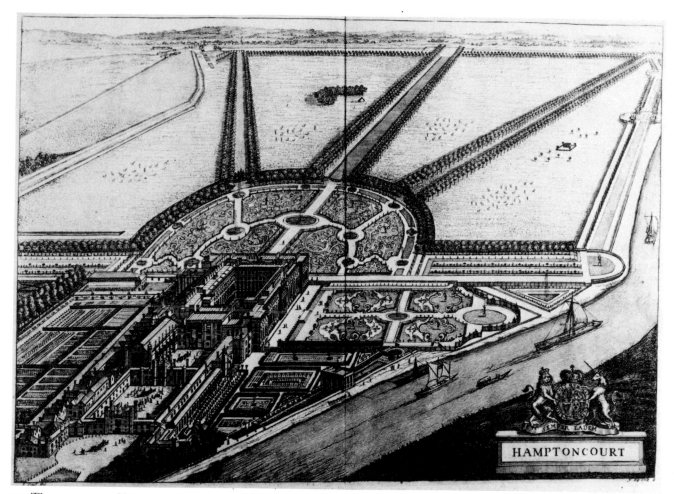

59. The avenues at Hampton Court extend the authority of the Palace out into the countryside. *Britannia illustrata*, 1709.

These thoughts all led far beyond the confines of the enclosed garden, and there was another tendency of a more obviously secular and political kind, making towards the same end. Renaissance princes added enclosure to enclosure, betraying an infinite appetite for the expansion of gardens of ostentation. In England, at least, this development received a setback when the great royal gardens were destroyed under the Commonwealth. But with the return of the Stuarts a counter-revolution set in. Most of the residences of the nobility and gentry lay in mild scenery and soft climes. Here, regular planting with long avenues, as recommended by André Mollet and practised in France,[44] would carry the influence of government out, beyond the boundaries of the formal garden, and into the countryside to subdue it (Plate 59). In Stuart times the public highway was envisaged, in poetry at any rate, as the haunt of outlaws and of rootless men: the avenue, on the other hand, represented an ordered, or, according to your point of view, a subjugated society. The avenue flattered its owner. It was a place where

> like a *Guard* on either side,
> The Trees before their *Lord* divide.[45]

93

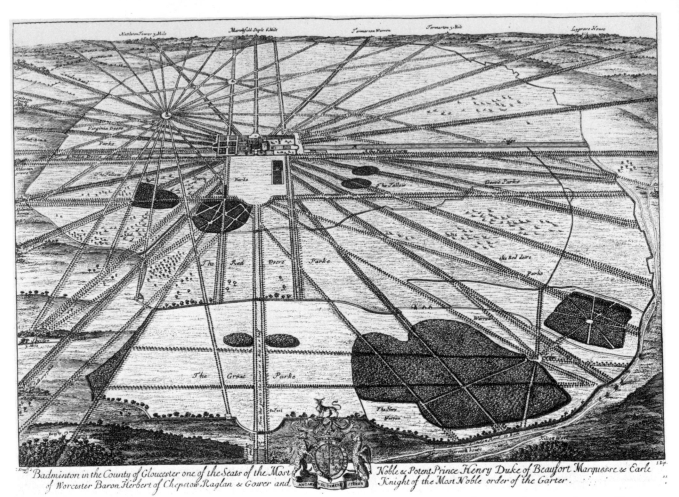

Badminton in the County of Gloucester one of the Seats of the Most Noble & Potent Prince Henry Duke of Beaufort Marquesse & Earle of Worcester Baron Herbert of Chepstow Raglan & Gower and Knight of the Most Noble order of the Garter.

60. At Badminton the avenues ran for miles and subordinated the surrounding villages to the great house. Contemporaries could justify such an arrangement as bringing the fallen world back into an hierarchical order.
Britannia illustrata, 1709.

By laying down avenues, a man could exhibit the extent of his estate. Whateley caught the feeling exactly when he wrote of 'the property of a riding *to extend the idea of a seat*, and appropriate a whole country to the mansion'.[46] It was characteristic of the late seventeenth century that however far an avenue extended it should proceed in a straight line from the house. Moses Cook spoke of setting out stakes to the front door,[47] and when this was done, as at Badminton (Plate 60), with its radiating avenues extending for miles to the horizon in every direction, the whole of a man's possessions could be surveyed from a single point. The exiled aristocracy had spent their time in France and learned their lesson well. The avenue was the policeman's truncheon of topography. Their hosts 'extended the dominion of art over nature uncompromisingly and without quarter',[48] and later in the century, when the Duke of Montagu returned from his embassy at Versailles, he projected an avenue which was to run in a direct line for seventy miles from his seat at Boughton in Northamptonshire to London.[49] Faced with these developments, and with the practice, as at Chatsworth (Plate 61), of extending the garden by adding one garden feature to another until the boundaries reached the very Peak itself, we

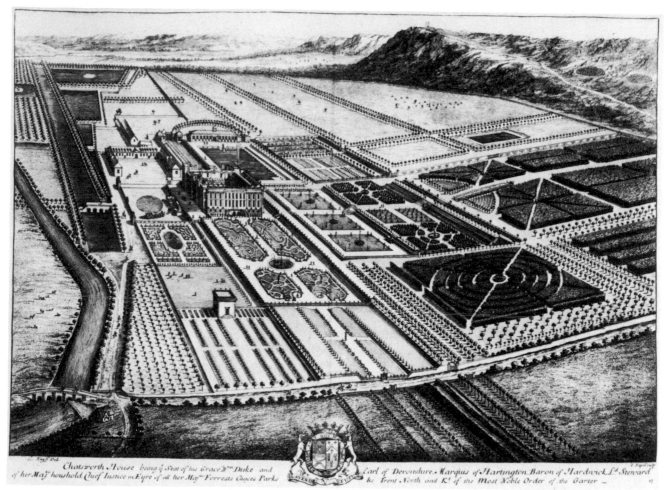

61. Chatsworth, where one formal garden was added to another until man's order and nature's wilderness clashed at the foot of the Peak.
Britannia illustrata, 1709.

feel ourselves to be in the presence both of autocracy, for if anything is out of place in an avenue or formal garden, the autocrat can spot it at once, and so too can the functionary whom he employs, and of the restless search for dominion which Hobbes attributed to mankind.

After the Glorious Revolution of 1688 the royalist counter revolution was transformed into the systematic aggrandizement of the Whig aristocracy. Garden writers caught their patrons' mood exactly: Le Blond urged them to avoid the mean and pitiful, and to seek the great and noble. Richard Bradley criticised English designers for not having dared to study what was great, and argued in favour of gardens of 'grandeur and recreation' – two concepts that it would not have been easy to reconcile a hundred, or even seventy-five years earlier.[50] Stephen Switzer objected to 'those crimping, diminutive and wretched Performances we everywhere meet with', and expressed his preference for 'large prolated Gardens and Plantations', and the extended gardening of what he (mis)called La Grand Manier.[51] Batty Langley communicated his ideas in a book intended to implant *New Principles of Gardening* (Plate 62) after a more GRAND manner than had

62. The Grand Manner.
   Batty Langley *New Principles of Gardening*,
   1728.

been done before.[52] Even the mild Addison preferred Versailles to Hampton Court or Chatsworth, because it was less mean-spirited.[53]

Between the optimists and the aggrandizers gardening passed into landscape, and as it did so the great sea-change in men's outlook towards nature got under way. The idea that nature was not totally corrupt was, as we have seen, implicit in the concept of reading in the book of God's works collected together into a Botanic Garden. But even those who doubted whether nature had been perverted at the Fall, often drew a distinction between the wild flowers, and the larger blossoms which were fit to be collected into a garden of re-creation, though even the wild flowers, Evelyn thought, might still have their uses, as guides to the crops that should be planted when land was reclaimed and put under cultivation.[54] And many even of those who looked kindly upon the plants and animals, admiring their beauty and fitness, continued to suppose that the physical world in which they were accommodated, with its seasons' difference and its ungainly topography, was deformed. Handel, for example, still followed Isaiah in looking forward to a day when every valley would be exalted and every mountain and hill be made low.[55]

This view had never passed unchallenged. In the early sixteenth century, Coverdale, the first man to translate the bible into English, said that God had not made the ground 'like on every side . . . And hereof have we the valleys, plains, mountains and hills'.[56] In 1627 Hakewill published a 'censure of the common error touching Natures perpetuall and universall decay'.[57] Milton included a rolling countryside in his description of the Garden of Eden, and the Garden itself, following Dante, was set on a mount. Now, in the second half of the seventeenth century the remaining barriers fell. Milton's rose had still been without a thorn until the Fall,[58] but Ray preferred to admire the functional virtues of prickles to enable plants to protect themselves against grazing animals. Ray leapt, too, to the defence of mountains. Drawing attention to the fact that 'Mountains have been lookt upon by some . . . as signs and proofs that the present Earth is nothing else but a heap of Rubbish and Ruins'. Ray praised them for their good providence in causing rivers to run, and for their grandeur, which literally led the mind to higher things.[59] In 1693 Beaumont attacked the very citadel of the lapsarians when he cast doubt upon the retributive origin of the seasons. To him, the four seasons were analogous to the four elements, the four humours in men's bodies, and the four quarters of the world, and they seemed to exhibit such a 'mutual Connexion and Dependence on each other for the general benefit of the Earth' that he was bound to conclude 'the four Seasons to have been from all Ages'.[60] This new outlook reached a wider public in the eighteenth century in Thomson's poem *The Seasons*. Thomson showed his awareness of the older tradition when he referred to 'the high-piled hills of fractured earth', and to the period when

> Great Spring, before,
> Greened all the year; and fruits and blossoms blushed,
> In social sweetness, on the self-same bough.

But these were fables. To him, Nature was a God

> who, with a master-hand,
> Hast the great whole into perfection touched.

The Seasons danced 'harmonious', and Richard Bradley drove the last nail into the coffin of the old system when he concluded that animals had, after all, been intended to hunt each other from the beginning.[61]

And so men's attitudes towards nature changed, and as they did so, gardeners, who had burst the bounds of the *hortus conclusus* either in an optimistic attempt to overcome the consequences of sin, or with intent to reduce a rebellious people to subjection, began to stress the idea that nature likes variety. Once again, this was not without precedent. Bellarmine had praised the multitude and variety of things, and Sir Matthew Hale had drawn attention to the way in which 'the Animals and Vegetables . . . contribute to the beauty, glory, ornament and variety of the whole' and make the world '*valde bonum*'.[62] Beaumont argued that 'an orderly vicissitude of things' was most pleasant to us,[63] and Ray that God had introduced variety into nature in order to demonstrate his power. Most flying creatures have feathers, because feathers pleased God best, but some creatures that fly do not possess feathers, and this shows that God could have filled the air with a different kind of songster had he wished.[64] Variety was God's choice; Le Blond said that everyone was now convinced that the greatest beauty of gardens lay in variety, Le Nôtre sought to supply 'relieving variety', and Timothy Nourse said that variety sweetened the mind with content.[65] This doctrine was now appealed to even in the case of religious dissent, and in 1681 Edward Pearse, in

A Conformists's Plea for the Nonconformists, argued that 'God, both in the works of Nature, Providence and Grace, is most glorious in Variety, in Multiformity'.[66]

Variety was supplied not simply by designing parterres of different kinds, and by including separate, specialised gardens within the framework of the whole, but by contrast. This had been an important element in the intellectual impact made by the *hortus conclusus*. Edmund Spenser had contrasted the 'daintie lineaments of beauty' with 'the rude thickets and craggie clifts',[67] and the poet Joseph Beaumont had said that men only learned to appreciate good things by being 'forc'd to confesse / A Gardens Blessing by a Wildernesse'.[68] An enclosed garden, even a *hortus conclusus*, should include a gate or *claire voie*, through which one could look out in safety upon the dangerous world outside. Now, as the gardens of the nobility became more and more enlarged, mottes (Plate 63) and platforms had to be constructed to enable their owners, some to survey their own properties, and others, the more sensitive, who knew that were all nature to be conquered, held down by avenues, and laid out in squares, the garden itself would lose its meaning, to appreciate the contrast between the extended garden and the world outside, whether as an object of fear, of curiosity, or as a means to aesthetic inspiration.

These contrasts could be drawn either between the cultivated garden of leisure and the everyday working world of agriculture outside, or between the garden as what was cultivated, and the uncultivated forest, heath or moor. The first found expression in Robert Castell's *Villas of the Ancients* published in 1728, and in the many attempts to reinterpret Horace's and Pliny's country houses in an English scene. According to this line of thought, the mansion and its garden were to be set in the centre of a scene of happy agricultural labourers, and commentators have rightly interpreted Bridgeman's invention of the ha ha, and Kent's famous view of the fields across the elbow of the river Cherwell at Rousham in these terms. But it was never generally so, and the dominant contrast adopted in the gardens of the Hanoverian succession was that between the garden and the wilderness. And since the real wilderness and the primitive life style described in *As You Like It* was now something that existed in the forests of America rather than the denuded commons and moors of England, the gardeners of the seventeenth and early eighteenth centuries were led to recover the intellectual stimuli it provided by the creation of artificial wildernesses within the confines of the gardens themselves.

The practice appears, in elaborate form, at Castle Howard (Plate 64), where the design for a thicket behind the house was labelled a wilderness. To us, this arrangement, with its still regular, if not actually rectangular pattern of paths laid out between walls of clipped and levelled hedges, appears as architectural and artificial as a building. But contemporaries would have found the use of massed planting dense enough to harbour wild animals, and the abandonment of straight paths, in following which it was impossible to get lost or even to pass out of sight, in favour of curves and recesses where a couple on foot might be ambushed, or could lose themselves and snatch a kiss, adventurous enough to justify the name it bore. This line of development was to be carried much further by Switzer, who spoke of 'forest', and Batty Langley who spoke of 'rural' gardening.[69] Both meant the same thing, dense groves of trees threaded by sinuous paths (Plate 65), and the upshot was that the formal garden of leisure with its parterres came to be separated from

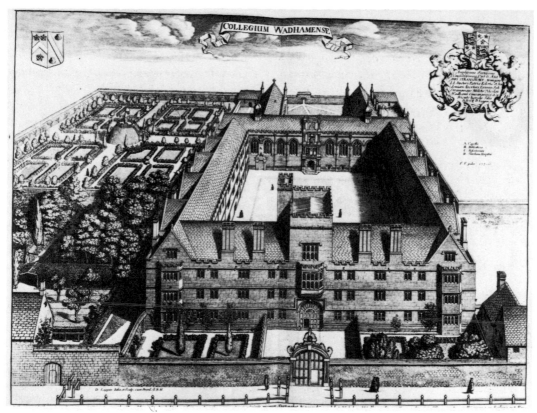

63. Loggan's view of Wadham College, Oxford, showing the motte in the centre of the garden.
Oxonia illustrata, 1675.

64. Castle Howard with the wilderness at the back of the house.
Colen Campbell *Vitruvius Britannicus*, 1725.

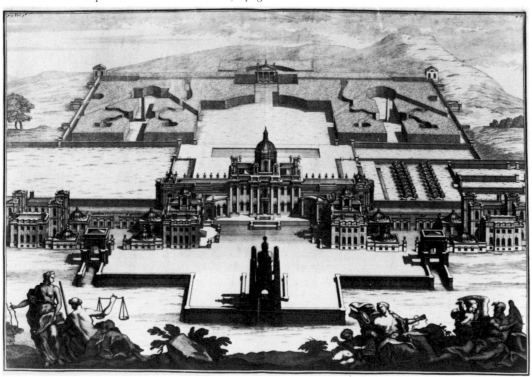

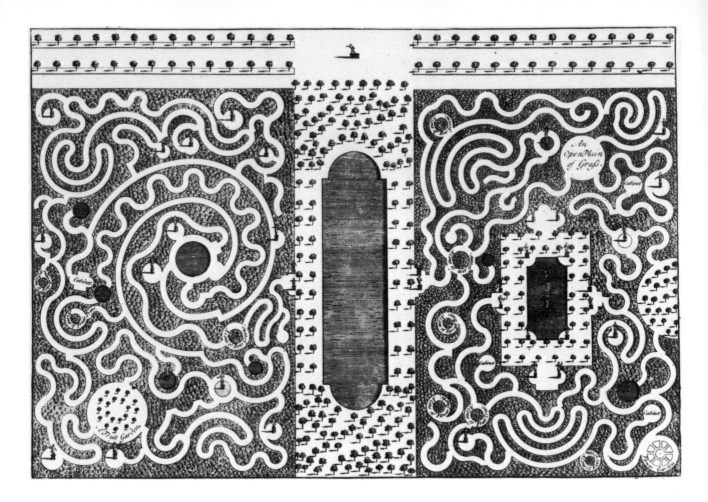

the working fields by a *cordon sanitaire* of make-believe plantations. The fanciful landowner could then pretend that the formal garden represented the re-created Garden of Eden, and the wilderness the primitive landscape of the golden age – thus getting the best of two traditions, or that the formal garden stood for Europe, and the wilderness for America, a concept made all the more vivid if, as in James's plan for a 'magnificent' garden (Plate 66) the two elements were separated by a canal. Above all, though, this re-arrangement of the pieces served the social demands of the day. The reconquered population, reduced to their subordinate ranks, could now be put out of sight altogether.

This variety, and these contrasts of the late seventeenth and early eighteenth century sprang out of the growing belief that, far from being corrupted, nature was innocent. But beliefs as deeply ingrained as the idea that the whole world had been corrupted at the Fall do not die all at once and in a single generation. It is tempting, therefore, to see something of the old idea of original sin still lingering in the distinction drawn in the middle of the eighteenth century between 'common' nature and her perfected form, and it is impossible not to be struck by the way in which, for all their tolerance of variety, gardeners still clung to the concept of order. This was apparent in the designs of the Renaissance gardeners, to whom straight lines and right angles were the marks of intelligence, and in the many attempts to present gardening in architectural terms. Sir William Temple de-

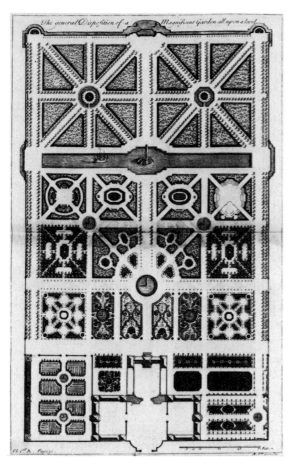

65. (facing page) Batty Langley's
serpentine wanderings.
New Principles of Gardening, 1728.

66. (left) A Magnificent Garden, with a
canal separating the formal from the
forest garden.
J. James *The Theory and Practice of
Gardening*, translated from the
French of A. J. B. Le Blond, 1712.

clared that 'Among us, the Beauty of Building and Planting is placed chiefly in some certain Proportions, Symmetries or Uniformities, our Walks and our Trees ranged so as to answer one another, and at exact Distances', and suggested that the various compartments of the garden should be 'like rooms out of which you step into another'.[70] Even in the twentieth century, commentators slip easily into describing the terraces of these late seventeenth-century gardens as outdoor rooms, and the walks which joined them as corridors.[71]

One after another the gardeners of this later period all sang the praises of order. Not everyone was as rigorous about it as John Laurence, who wanted all his hedges to be 'straight, and regular', and his trees to stand in rank and file 'with Order', 'Improv'd by Discipline, like martial Band',[72] or as ambitious as Beverley, who seems to have ruminated on the possibility of shearing mountains off into regular pyramids, and arranging them in rows or quincunx.[73] But Le Nôtre sought relieving variety combined with overall uniformity,[74] Bradley while repeating that nature was full of variety, thought that the task of the gardener was to reconcile Rule and Liberty,[75] and even the third Earl of Shaftesbury, who said that if forced to choose between order and chaos he would prefer 'all the horrid graces of the *Wilderness* itself, as representing NATURE more' to 'the mockery of princely gardens', admitted that God had made men 'as ardently desireous of the Beauty of *Order*, as of any Indulgence of Sense'.[76] The same theme is still apparent in the slightly more relaxed attitude expressed in Alexander Pope's ideal of order in variety, pleasing confusion, *discordia concors*,[77] and in William Gilpin's praise for the gardens at Stow,

> Here Order in Variety we see,
> Where all things differ, yet where all agree.[78]

The result was that the so-called irregular gardens of the first thirty years of the eighteenth century were still, in some ways, as formal as Versailles. The gardens of this period were formed along strong axial vistas.[79] Nature, according to Pope, was to be 'methodis'd',[80] and this was the feeling that informed the work of Bridgeman, the greatest of the gardeners of this period. In order to expand the garden, Bridgeman pulled down the walls which were still being built as late as Vanbrugh, banished topiary, and began to let the foliage go free in order to provide variety (Plate 67). But he still kept a tight and artificial grip upon the whole design, imposing order by means of the straight walks lined by clipped hedges with which he enclosed his loose groves and wildernesses. In this manner, while admitting nature's infinite variety into mankind's order, he still saw his primary function as being that of imposing mankind's order upon nature's variety, and this comes out very clearly in his plan for the gardens at Rousham, with its long straight axis, and its canalised right hand bank to the river Cherwell.[81]

It took a century or more, then, to escape from the idea of an earth disordered and poisoned as a result of the Fall, and to arrive at the informal dogma of the fluent curve exemplified in the work of Capability Brown in the second half of the eighteenth century. Brown swept away the old formal gardens, and tidied flowers and fruit away into an enclosure that was carefully sited out of view of the house. He cut down the avenues planted by his predecessors, and replaced them with circumferential belts and clumps of trees. He respected the natural rise and fall of the ground, and improved upon it where necessary in order to create a lake. Finally, he laid turf, and brought in herds of cattle and deer to graze the 'lawns' right up to the front of the house. Animals were thus, for the first time, brought to the centre of the garden stage.

Genesis made it clear that in the original garden man had lived at peace with the animals, and this idea was reflected in the work of the late sixteenth- and seventeenth-century poets. Du Bartas said that in the Garden of Eden

> The fiercest Beasts, would at his word, or beck,
> Bow to his yoake their self-obedient neck,
> As now the readie Horse is at command.[82]

Milton, whose botanical descriptions were not, perhaps, the strongest element in his great epic, rose to this theme, and to the way in which

> All beasts of the earth, since wild, and of all chase
> In wood or wilderness, forest or den,

lived at peace among themselves and with men.

67. Order and variety at Hall Barn. Copyright *Country Life*.

> Sporting the lion ramped, and in his paw
> Dandled the kid; bears, tigers, ounces, pards,
> Gambolled before them . . .[83]

It seems to have been this passage which sprang to Horace Walpole's mind when looking at a Capability Brown landscape. Walpole saw in Milton's description of the Garden of Eden 'a warmer and more just picture of the present style' than could have been painted by a landscape artist, even by Claude Lorrain himself.[84] Hence his famous quatrain.

> With one lost Paradise the name
> Of our first ancestor is stained;
> Brown shall enjoy unsullied fame
> For many a Paradise regained.[85]

Walpole was not suggesting that Brown had read *Paradise Lost*, though there was no reason why a gardener of humble origin should not have done. He was drawing attention to the fact that animals, which had been associated with the wilderness and even with the devil in the middle ages, and confined to cramped pens by well-

68. A 'Capability' Brown garden; Southill, with horses, cattle and deer grazing right up to the house. W. Watts *The Seats of the Nobility and Gentry*, 1779.

meaning but ignorant keepers in the early menageries and Zoological Gardens of the Renaissance, were now being allowed to roam freely, browsing the foliage of the trees, as they wandered among the flowing and ever receding curves of a gentle landscape. Artists had always preferred to represent the Garden of Eden as a place where plants flourished and animals sat or strolled among them, as the paintings by Brueghel in the Victoria and Albert Museum (Col. Pl. I), by Van Eyck and Rubens in the Mauritshuis, and of Roelandt Savery in Prague (Plate 70) show. To the practical gardener, charged with the responsibility of realising this Paradise, plants and animals had always been, in the last resort, irreconcilable. However much he would have liked to bring the two together, animals had, in fact, to be excluded from the Botanic Gardens, or housed in separate collections. Now, the wheel had turned, and with Brown it was the garden plants which were excluded from the ideal landscape and relegated to an isolated enclosure of their own, and the animals which were made the focus of attention.

Walpole's opinions command respect. But there was at least one reason why even he must have realised that, in the second half of the eighteenth century, it was more than a little fanciful to describe an English landscape garden in seventeenth-century terms of a re-created Garden of Eden. This is that it was difficult to sustain the equation of the horses and cattle, and the timid deer in Ashburnham Park and

69. Not a Brown garden, but a garden in the Brown manner: Copped Hall.
W. Watts *The Seats of the Nobility and Gentry*, 1779.

Southill (Plates 68, 69) with the 'bears, tigers, ounces, and pards' of Milton's epic. Brown's gardening was all the rage, but some of Walpole's contemporaries criticised it for its tameness,[86] and Brown's clients' narrowly selective choice of animals was not the least among its many 'tame' features. To this one might add Oliver Goldsmith's criticism that, however idealistically conceived, English landscape gardening involved social costs.[87] The construction of a park necessitated large schemes of social engineering. Whole villages were removed, as at Nuneham Courtenay, and roads diverted as at Highclere, in order that the beneficiaries of the revolution of 1688 might be able to surround themselves with docile animals rather than vulgar and sometimes discontented men. Everything that was rude or rowdy was dumped outside the park gate, and when, in due course, the rising tide of interest in animals found expression, in 1826, in the formation of the London Zoo, every care was taken 'to prevent the contamination' of the garden in Regent's Park 'by the admission of the poorer classes of Society.'[88] The Zoo in the park could, indeed, have been a Garden of Eden, but in other ways, too, the objective was not idealistic. The original prospectus spoke of collecting animals 'which may be judged capable of application to purposes of utility',[89] for food, burden, or service, and even today its descendant, the modern wild-life park, is probably esteemed more as a refuge for endangered species than for its Edenic qualities.

70. *Paradise* by Roelandt Savery (1576–1639) is perhaps the most luxuriant painting of its kind. National Gallery, Prague.

71. Edward Hicks (1780–1849) painted many variations upon *The Peaceable Kingdom*. This one dates from *c.* 1830–40, and links the original contract with the Indians, made when Pennsylvania was founded in 1681, to the Edenic aspirations still prevalent among U.S. settlers early in the nineteenth century. Worcester Art Museum, Worcester, Mass.

There were, it is true, exceptions to the tameness of Brown's parks. In France, Verniquet proposed a scheme for displaying animals in a series of gardens resembling their natural environments, attached to the Jardin des Plantes.[90] In England, the cheetah, brought from Madras and presented to George III, was released in Windsor Park and immortalised in a painting by Stubbs.[91] Early in the nineteenth century, Charles Waterton turned Walton Hall, Yorkshire, into the first genuine wild-life park.[92] But the dated ideal of peace among the animals, expressed in *Isaiah*, has remained the preserve of poets and visionaries. In Pennsylvania, Edward Hicks interpreted the wolf, the leopard, the bear, and the lion, as figures of bad, melancholy, sanguine, phlegmatic, and choleric men and women, and the lamb, the kid, the cow, and the ox, as emblems of good ones.[93] Some authorities suppose that Hicks may even have been acquainted with engravings of Roelandt Savery's paintings of Paradise.[94] However that may be, the juxtaposition of good and bad animals in Hicks's famous 'peaceable kingdom' paintings (Plate 71), stands for the resolution of all the conflicts and tensions of earthly life. In the twentieth century, the same thought has found expression in the engravings of Fritz Eichenberg.[95]

Today, the whole re-creational ideal seems unattainable. But it was not always so. In the sixteenth and seventeenth centuries re-creation was something to strive for, not sigh for. Admittedly there were many stumbling blocks and unresolved contradictions in the way. Worldly men cultivated anti-Paradises, or gardens of physical delights. Spiritual men enveloped the subject in magical overtones, and in some cases, like Culpeper, with a belief in astrology, which I have excluded from this study.[96] It was always much easier to think in terms of recreating the vegetable side of the original Garden of Eden, than it was to think of the animal. Even in the Botanic Garden itself there was a curious tug-of-war between the botanist's ambition to endow plants with names in Latin, and the theologian's insistence upon Hebrew as the language of Paradise. Yet more fundamental still was the implication, lying behind the attempt to recover the original Garden of Eden, and to unlock the botanical secrets of health, medicine, and life, that there were alternative routes open to mankind which cut in upon the doctrine of the atonement. The Christian had to be wary when he lifted his eyes from the cross to the creation. The cynic may say that these contradictions never had to be resolved, because the whole object of re-creational enterprise was unattainable from the start. Others, who resonate more easily to the wavelengths upon which our ancestors communicated, may feel more sympathy with a sometimes ill-defined, but very pervasive hope for a better world.

NOTES TO THE TEXT

The place of publication is London, unless otherwise stated.

ABBREVIATIONS

DWW   Saluste du Bartas *His divine weekes and workes* translated by Joshuah Sylvester, 1605.

PL   *Paradise Lost, The poems of John Milton* edited by John Carey and Alastair Fowler, 1968.

PM   Abraham Cowley *Plantarum, the third and last volume of the works of Mr Abraham Cowley, including his six books of plants* 1721.

SB   P. Stephens and W. Brown *Catalogus Horti Botanici Oxoniensis* 1658.

EB   John Evelyn 'Elysium Britannicum', MS in Christ Church Library, Oxford.

EJB   'Vertumnus' *An epistle to Mr Jacob Bobart* 1713.

INTRODUCTION

1. R. T. Günther *Oxford gardens* Oxford 1912, p. 188, from the account by Thomas Baskerville, *c.* 1683.
2. Arnold Williams *The common expositor. An account of the commentaries on Genesis 1527–1633*, Chapel Hill, 1948.

CHAPTER I

1. E. V. Rieu tr. *The Odyssey* Harmondsworth 1946, p. 115; *Isaiah* XI:6; Virgil, *Fourth Eclogue.*
2. *Genesis* I:30.
3. *ibid* I:29, II:19, I:28.
4. *ibid* II:8–14.
5. *ibid* II:9.
6. Charles Stengel *Hortensius, et Dea Flora, cum Pomona historice, tropologice, et anagogice descripti* Augsburg 1647, pt. I ch. IV.
7. *Genesis* II:15.
8. *ibid* I:27, II:21,22.
9. Tertullian, Basil, in George Boas *Essays on primitivism and related ideas in the middle ages* Baltimore 1948, pp. 17, 31–2. Jerome in G. H. Williams *Wilderness and paradise in christian thought* New York 1962, p. 43.
10. Boas *op. cit.* p. 18, Williams *op. cit.* p. 43.
11. Boas *op. cit.* pp. 43–44, see too Williams *op. cit.* p. 49.

12. Theophilus and Lactantius in Boas *op. cit.* pp. 16, 33–4.
13. *Genesis* III:16, 18, 19.
14. *ibid* II:17.
15. *Romans* VIII:22.
16. Augustine, Ernaldus, Honoré d'Autun in Boas *op. cit.* pp. 49, 74, 78.
17. Discussed by Honoré d'Autun and Hildebert, closely following Augustine in Boas *op. cit.* pp. 78, 81.

CHAPTER II

1. Theophilus, Basil, Lactantius, Ambrose in Boas *op. cit.* pp. 15 seq., 31, 37, 44–5.
2. A. Bartlett Giamatti *The earthly paradise and the renaissance epic* Princeton 1966, p. 12. P. Ildefonse Ayer 'Où plaça-t-on le paradis terrestre?' *Études Franciscaines* XXXVI 1924, pp. 119–21.
3. *Genesis* V:24. 2 *Kings* II:11.
4. *Luke* XXIII:43. Ayer *op. cit.* p. 126.
5. St Paul in *Acts* IX:3, 2 *Corinthians* XII:2–4.
6. Tertullian in Ayer *op. cit.* p. 123.
7. M. M. Innes tr. *The Metamorphoses of Ovid* Harmondsworth 1955, pp. 33–4.
8. Giamatti *op. cit.* p. 68.
9. *Fourth Eclogue.*
10. C. L. Sanford *The quest for paradise, Europe and the American moral imagination* Urbana 1961, pp. 8–9.
11. M. L. Gothein *Geschichte der gartenkunst* 2 vols Jena

1914, I 172. Teresa McLean *Medieval English gardens* 1981, pp. 16, 18, 47–9, 123.

12. Stanley Stewart *The enclosed garden, the tradition and the image in seventeenth-century poetry* Madison 1966, pp. 52, 147.
13. Williams *op. cit.* pp. 39, 43.
14. *ibid* pp. 63–4.
15. *Song of Solomon* IV:12.
16. Stewart *op. cit.* pp. 62, 29–30.
17. McLean *op. cit.* p. 99.
18. *ibid* p. 133.
19. Stewart *op. cit.* p. 116.
20. Williams *op. cit.* ch. 1.
21. 1 *Kings* XIX:8, 2 *Kings* IV:38.
22. *Isaiah* XL:3.
23. *Matthew* III:3, *Mark* I:3, *Luke* III:4, *John* I:23.
24. *Hosea* II:18.
25. Williams *op. cit.* p. 28.
26. Giamatti *op. cit.* pp. 81–3.
27. Williams *op. cit.* pp. 42–3.
28. *Purgatorio* XXVIII, *Inferno* I.
29. Giamatti *op. cit.* p. 136.

CHAPTER III

1. See Erasmus Warren *Geologia: or, a discourse concerning the earth before the deluge* . . . 1690, p. 271.
2. H. R. Patch *The other world, according to descriptions in medieval literature*, Cambridge, Mass. 1950, p. 143.
3. Henry Hare *The situation of paradise found out* . . . 1683 pp. 20–24.
4. Stewart *op. cit.* pp. 78–9.
5. Sir Walter Raleigh *History of the world* 1614, Book I, ch. III para 2. Stengel *op. cit.* pt. I ch. II, p. 28.
6. Stengel *op. cit.* pt. I ch. II, p. 29.
7. H. Baudet *Paradise on earth, some thoughts on European images of non-European man* New Haven 1965, p. 15.
8. Stengel *op. cit.* pt. I ch. II, p. 30. Raleigh *op. cit.* Book I ch. III, para 2.
9. Salomon van Til *Dissertationes philologico-theologicae* Leyden 1719, liber tertius, caput VI.
10. *PL* IV:158–63.
11. Stengel *op. cit.* pt. I ch. II, p. 28.
12. Giamatti *op. cit.* p. 79.
13. Ayer *op. cit.* p. 117 seq.
14. For early maps see Konrad Miller *Mappaemundi. Die ältesten weltkarten* Stuttgart 1895.
15. S. E. Morison *Admiral of the ocean sea* 2 vols, Boston 1942, I:121.
16. M. Éliade 'The yearning for Paradise in primitive tradition', *Daedalus* 88 1959, p. 261.
17. Raleigh *op. cit.* Book I ch. III, para 3, p. 26.
18. M. W. Labarge *A baronial household of the thirteenth century* 1965, p. 88.
19. *Acts* VIII:28.
20. Baudet *op. cit.* p. 15.
21. *idem.*
22. C. R. Boxer *The Portuguese Seaborne Empire, 1415–1825* 1969, p. 27.
23. Baudet *op. cit.* pp. 23–4.

24. Sergio Buarque de Holanda *Visão do paráiso* second edn Sao Paulo 1969, pp. 153, 208.
25. Morison *op. cit.* I:124.
26. *ibid* I:124–5.
27. Baudet *op. cit.* p. 26, following Morison *op. cit.* I:187.
28. Buarque de Holanda *op. cit.* p. 179 fn. 2.
29. Morison *op. cit.* I:428 n. 9.
30. *ibid* II:283.
31. *ibid* II:284–5.
32. J. H. Elliott *The old world and the new, 1492–1650* Cambridge 1970, p. 20.
33. H. Levin *The myth of the golden age in the renaissance* Bloomington 1969, p. 65.
34. Buarque de Holanda *op. cit.* p. 239.
35. *ibid* 276.
36. *ibid* 134.
37. *ibid* 295.
38. *ibid* 35.
39. J. L. Phelan *The millenial kingdom of the Franciscans in the new world* Berkeley and Los Angeles 1956, p. 22.
40. Elliott *op. cit.* p. 24. Phelan *op. cit.* pp. 24–5.
41. See especially A. W. Crosby *The Columbian exchange, biological and cultural consequences of 1492* Westport Conn. 1972.
42. Phelan *op. cit.* p. 25.
43. Crosby *op. cit.* pp. 12–13.
44. Quoted in Levin *op. cit.* p. 66.
45. *ibid* p. 184.
46. Phelan *op. cit.* p. 72.
47. Raleigh *op. cit.* Book I ch. III, paras 8, 15.
48. *DWW* p. 277.
49. pt. I ch. V.
50. *PL* XI:831–5.
51. Marmaduke Carver *A discourse of the terrestrial paradise* 1666, 'To the Reader'.
52. Thomas Park ed. *The Poetical works of Nathaniel Cotton* 1806, p. 126.

CHAPTER IV

1. Elliott *op. cit.*
2. Quoted in Crosby *op. cit.* p. 9.
3. L. P. Wilkinson *Horace and his lyric poetry* Cambridge 1945, pp. 165–6.
4. Allen G. Debus *Science and education in the seventeenth century: the Webster–Ward debate* 1970, pp. 7–9.
5. Ralph Austen *The spiritual use of an orchard or garden of fruit trees* second edition 1657, repr. 1847. p. xiv.
6. S. du Bartas *Part of du Bartas, English and French* translated by William L'Isle, 1625, pp. 116–17, from *The third Booke of Noe.*
7. *DWW* p. 6.
8. *PL* VIII:67–8.
9. Maren-Sofie Røstvig *The happy man* second edn 2 vols, Oslo 1962, I:127.
10. Stewart *op. cit.* p. 116.
11. William Coles *Adam in Eden: or, Nature's paradise* 1657, 'To William Coles'.

12. *DWW* p. 7.
13. See A. G. Hodgkiss *Discovering antique maps* Aylesbury 1977, p. 16.
14. Baudet *op. cit.* p. 17.
15. *PM* pp. 393, 418–19.
16. Julia S. Berrall *The garden* Harmondsworth 1978, p. 119.
17. Reproduced in C. R. Boxer *The Dutch seaborne empire* 1965, frontispiece.
18. *PL* V:338–9.
19. John Rea *Flora: seu de florum cultura* 1665, 'To the same lady'.
20. Jacob Burckhardt *The civilisation of the renaissance in Italy* London 1945, pp. 175–6.
21. B. D. Jackson 'A draft of a letter by John Gerard', *Cambridge antiquarian communications, Vol. IV 1876–80* 1881.
22. G. Porro *L'Horto de i semplici di Padova* 1591, poem to Cortusius; P. Hermann *Paradisi Batavi* Leyden 1689, to the reader; P. Vallet *Le jardin du roy* Paris 1608, G. Desdames Rothomag.
23. R. T. Günther *op. cit.* p. 189.
24. *SB* 'The Preface to the Philobotanick Reader'.
25. *EJB* p. 13.
26. Stengel *op. cit.* pt. I ch. X, XI.
27. R. J. W. Evans *Rudolf II and his world* Oxford 1973, p. 177.
28. Edmund Spenser *The Faerie Queene* Book III cant. VI, v. 35.
29. Porro *op. cit.* 'Ai studiosi lettori'. C. Daubeny *Oxford botanic garden; or a popular guide to the botanic garden of Oxford* 1853. J. Hill *An idea of a botanical garden in England* 1758, pp. 13–14.
30. Henry Hawkins *Partheneia sacra* 1633, p. 6.
31. *PM* p. 387.
32. Roy Strong *The Renaissance garden in England* 1979.
33. John Evelyn *Kalendarium* 1664, introduction.
34. EB pt. I ch. I.
35. *idem.*
36. *ibid* pt. II ch. XVII.
37. *idem.* For a description of the garden at Paris see Guy de la Brosse *Description du Jardin Royal des plantes médecinales estably par le roy Louis le juste à Paris* 1636, pp. 11–26.
38. C. R. Boxer 'The Portuguese in the east 1500–1800' in H. V. Livermore ed. *Portugal and Brazil* Oxford 1953, p. 217.
39. P. Kolb *The present state of the Cape of Good Hope* 1738 edn. I, pp. 353–4.
40. Guy Tachard *Voyage de Siam des pères Jésuites* 2 vols, Paris, 1686, II p. 72. Mia C. Karsten *The old company's garden at the Cape and its superintendents* Cape Town 1951, p. 111.
41. Sir William Temple *Works* 2 vols 1720, I pp. 186–7.
42. Karsten *op. cit.* p. 114.
43. Wilfrid Blunt *The compleat naturalist, a life of Linnaeus* 1971, pp. 103, 117.
44. EB pt. II ch. XIII, p. 236. Røstvig *op cit.* I p. 160.
45. See A. O. Lovejoy *The great chain of being* Cambridge, Mass. 1936.
46. P. A. Robin *Animal lore in English literature* 1932, p. 32.
47. *DWW* p. 290.
48. *Works* ed. J. Spedding, R. L. Ellis and D. D. Heath, 14 vols, 1857–74, II p. 531.
49. p. 1618.
50. John Jonstone *An History of the wonderful things of nature* 1657 edn. p. 131.
51. Porro *op. cit.* 'Ai studiosi lettori'.
52. Hugh Honour *The new golden land, European images of America from the discoveries to the present time* 1976, p. 35.
53. Petri Pawi *Hortus publicus academiae Lugduno-Batavae* 1601.
54. Martin Welch *The Tradescants and the foundation of the Ashmolean museum* Oxford 1978, p. 6.
55. Blunt *Linnaeus* p. 95.
56. Boas *op. cit.* p. 74.
57. H. F. Clark 'Eighteenth century elysiums' in *England and the Mediterranean tradition* Oxford 1945.
58. EB pt. II ch. XIII.
59. Blunt *Linnaeus* p. 151.
60. D. E. Allen *The naturalist in Britain, a social history* Harmondsworth 1976, p. 29.
61. Francis Quarles *Emblemes* 1635, Book V no. 3, p. 252.
62. Boas *op. cit.* pp. 55, 68, 167.
63. *DWW* p. 198.
64. John Parkinson *Paradisi in sole, paradisus terrestris* 1629, 'To the courteous reader'.
65. EB pt. I ch. I.
66. Charles Webster *The great instauration* 1975, p. 468.
67. Stephen Blake *The compleat gardeners practice* 1664, p. 151.
68. Røstvig *op. cit.* II p. 45, Anon. 1751.
69. Parkinson *Paradisi in sole*, 'To the courteous reader'.
70. C. E. Raven *English naturalists from Neckham to Ray* Cambridge 1947, p. 234.
71. Porro *op. cit.*, poem to Cortusius.
72. *SB* R.I. to.
73. *EJB* p. 15.

Chapter V

1. *Ezekiel* XLVII 12.
2. Porro *op. cit.*
3. Richard Bradley *A philosophical account of the works of nature endeavouring to set forth the several gradations remarkable in the mineral, vegetable and animal parts of the creation, tending to the composition of a scale of life* 1721, p. 187.
4. de la Brosse *op. cit.* p. 19.
5. B. D. Jackson *op. cit.*
6. R. T. Günther *op. cit.* pp. 1–2.
7. *SB.*
8. EB pt. I ch. II, pt. II ch. XVI, p. 321.
9. Coles *op. cit.*, G. Wharton to W. Coles.
10. Porro *op. cit.* poem to Cortusius.
11. *SB.*
12. Coles *op. cit.*, 'To the Reader'.
13. *SB.*

14. Jonstone *op. cit.* p. 128.
15. Joseph Fletcher *The perfect-cursed-blessed man* 1628, p. 27.
16. *DWW* p. 347.
17. *ibid* p. 97.
18. *ibid* pp. 98–101.
19. *PM* pp. 282, 275, 254, 247–8, 248–9.
20. *ibid* p. 279.
21. L. G. Matthews 'Herbals and formularies' in F. N. L. Poynter ed. *The evolution of pharmacy in Britain* 1965.
22. John Hill *The useful family herbal* 1754, pp. x–xi.
23. e.g. Bartholomeus Ambrosinus *Panacea ex herbis quae a sanctis denominantur cocinnata opus* Bononiae 1630.
24. Robert Turner in Raven *English naturalists* p. 234.
25. B. D. Jackson *op. cit.*
26. Coles *op. cit.*, E. Philips to W. Coles.
27. *ibid* p. 3.
28. *PM* p. 245.
29. Jonstone *op. cit.* p. 128. Porta was credited with having founded, or re-founded the science.
30. Porta, in Agnes Arber *Herbals; their origin and evolution, a chapter in the history of botany, 1470–1670* Cambridge 1912, p. 208.
31. R. W. Crausius *Dissertatio inauguralis medica de signaturis vegetabilium* Jena 1697, p. 16.
32. Coles *op. cit.*, G. Wharton to W. Coles. Crausio *op. cit.* p. 17.
33. Coles *op. cit.* A.B. to W. Coles.
34. *ibid*, G. Wharton to W. Coles.
35. Arber *op. cit.* pp. 211–12.
36. L. G. Matthews *op. cit.* p. 193.
37. Arber *op. cit.* p. 211, based on Coles' 'Further account . . . of the method'.
38. Crausius *op. cit.* ch. III.
39. Jonstone *op. cit.* p. 131.
40. *EJB* p. 15.
41. C. E. Raven *John Ray, naturalist* 1950 edn. pp. 98–9, 455.
42. Coles *op. cit.* 'To the Reader'.

CHAPTER VI

1. Book III cant. VI, v. 33.
2. *DWW* pp. 273–6.
3. lines 13–14.
4. *PM* p. 387.
5. *PL* IV:155, and IV:148–9.
6. *ibid* V:323.
7. *ibid* V:394–5.
8. Habington *Poems* ed. K. Allott, 1948, p. 17.
9. Hare *op. cit.* p. 12.
10. Thomas Burnet *The theory of the earth* 2 vols 1684, I p. 178.
11. Francis Bacon *Of gardens* Ashendene Press 1897, p. 19. EB pt. II ch. XIV, p. 259.
12. *Pomona* 1664, p. 3.
13. EB pt. II ch. XIV, pp. 259–60.
14. Levin *op. cit.* pp. 86, 96.
15. Hawkins *op. cit.* pp. 136–7.

16. See Williams *op. cit.* pp. 49–50.
17. *EJB* p. 12.
18. vol. I p. 68.
19. Parkinson *Paradisi in sole* p. 8.
20. *EJB* p. 12.
21. James Hervey *Reflections on a flower garden* 1746, p. 64.
22. McLean *op. cit.* p. 52.
23. *ibid* pp. 25–6.
24. EB pt. II ch. XIV. Parkinson *Paradisi in sole* p. 8.
25. EB pt. I ch. I. For Marvell see Stewart *op. cit.* p. 183. Joseph Heely *Letters on the beauties of Hagley, Envil and the Leasowes* 1777, letter VII, pp. 152–3.
26. Raleigh *op. cit.* Book I ch. III, para 3 p. 37. Parkinson *Paradisi in sole* 'To the courteous Reader'. Sir Hugh Plat *The Garden of Eden, or, an accurate description of all flowers and fruits now growing in England* 1653, p. 31. William Lawson *A new orchard and garden* 1683, p. 53, 'What was Paradise? but a Garden, an Orchard of Trees and Herbs . . .'
27. *Song of Solomon* II:3, VII:8, VIII:5.
28. Stewart *op. cit.* p. 145.
29. E. Mâle *The gothic image* London 1961, pp. 106–7.
30. Robert B. Hinman *Abraham Cowley's world of order* Cambridge, Mass. 1960, p. 289.
31. Ralph Austen *A dialogue, or familiar discourse and conference between the Husbandman, and Fruit Trees, in his nurseries, orchards, and gardens* 1676, p. 1.
32. *ibid* p. 16.
33. *DWW* p. 330.
34. *PL* XI 185–9.
35. *EJB* p. 16.
36. Phelan *op. cit.* p. 57.
37. Christopher Hill *Puritanism and revolution* 1958, ch. XI.
38. Thomas Tryon *The way to health, long life and happiness* 1683, p. 79.
39. *idem.*
40. R. Dodsley ed. *The works of William Shenstone* 3 vols, fifth edn. 1777, II 133.
41. *To Saxham* lines 23–6, quoted in J. G. Turner *Topographia and the topographical poem in English 1640–1660* D. Phil. thesis presented to Oxford University 1976, from T. Carew *Poems* 1642, p. 46.
42. Røstvig *op cit.* I p. 29.
43. Levin *op. cit.* p. 172.
44. Günther *op. cit.* p. 189.
45. *Works* 2 vols 1720, I p. 189.
46. *Pomona* 1664, p. 1.
47. *ibid* p. 3.
48. *ibid* p. 2.
49. Mea Allan *The Tradescants: their plants, gardens and museum 1570–1662* 1964, pp. 37, 49.
50. Bodleian Library, Ashmole Ms. 1461.
51. *Works* 2 vols 1720, I p. 181.
52. Appendices i–iv to *The French gardiner* 1658.
53. McLean *op. cit.* p. 228.
54. Gothein *op. cit.* I p. 196.
55. Leonard Meager *The English gardner* 1699 edn, supplement, p. 133.
56. *PM* p. 396.

57. See J. Commelin *The Belgick or Netherlandish hesperides* 1683.
58. Temple *Works* 2 vols 1720, I p. 181. E. S. Rohde *The story of the garden* 1932, p. 149.
59. Allan *op. cit.* p. 144. Honour *op. cit.* p. 46.
60. Hervey *op. cit.* p. 88.
61. Hinman *op. cit.* p. 289.
62. See Hawkins *op. cit.* pp. 151–2, and Wither *op. cit.* Book III no. 38.
63. Raleigh *op. cit.* Book I ch. III, para 12 p. 56, and para 15 p. 64.
64. *DWW* p. 103, Raven *English Naturalists* p. 235.
65. *PM* p. 414.
66. Blunt *Linnaeus* p. 108.
67. Boas *op. cit.* pp. 43, 70.
68. Røstvig *op cit.* I pp. 272–3.
69. Renato Poggioli 'The pastoral of the self', *Daedalus* 88 1959, p. 695.
70. Allan *op. cit.* Appendix II, Catalogus Plantarum in Horto Johannis Tredescanti nascentium p. 276. *SB.* Guy de la Brosse *De la nature, vertu et utilité des plantes* 1628, Book I.
71. Hawkins *op. cit.* p. 156. *PM* p. 401. Raven *John Ray* p. 235.
72. Hawkins *op. cit.* pp. 151–8.
73. Arber *op. cit.* p. 5.
74. Jonstone *op. cit.* pp. 150–1, 127.
75. *The garden* stanza VIII.
76. *PM* p. 287.
77. *ibid* p. 321.
78. Røstvig *op. cit.* I p. 171.
79. They are not bent towards the earth, see F. E. Robbins *The hexaemeral literature, a study of the Greek and Latin commentaries on Genesis* Chicago 1912, p. 43.
80. Leo Spitzer 'Classical and christian ideas of world harmony', *Traditio* 2, 1944, p. 459; E. L. Jones 'The bird pests of British agriculture in recent centuries', *Agricultural history review* 1972, pp. 107–25.
81. *DWW* p. 275. Lawson quoted in Rohde *op. cit.* p. 128. *PL* IV 264. Evelyn *Pomona* p. 3, and EB pt. II ch. XIII.
82. Joseph Addison 'The pleasure of a garden' in A. H. Hyatt ed. *A book of old-world gardens* 1911, p. 15.
83. Spitzer *op. cit.* p. 457.
84. Austen *Spiritual use* proposition 93.
85. *ibid* prop. 1, p. 1.
86. *ibid* prop. 3, p. 7.
87. *ibid* prop. 7, p. 42.
88. *ibid* prop. 11, p. 58.
89. *ibid* prop. 31, pp. 162–3.
90. *ibid* prop. 29, p. 149.
91. *ibid* prop. 48, p. 233.
92. *ibid* prop. 59, pp. 278–9.
93. *ibid* prop. 64, p. 298.
94. *ibid* prop. 19, p. 92.
95. *ibid* prop. 66, p. 306.
96. Sanford *op. cit.* p. 40.
97. Austen *Spiritual use* prop. 66, p. 309.

Chapter VII

1. *DWW* p. 288.
2. *PM* p. 425.
3. *DWW* pp. 327–33.
4. Fletcher *op. cit.* pp. 26–7.
5. *DWW* p. 331.
6. Andrew Marvell *Upon Appleton house* stanza 96.
7. A. Brome *Songs and other poems* 1661, p. 159.
8. Røstvig *op. cit.* I p. 157.
9. Stewart *op. cit.* pp. 50–1.
10. Williams *op. cit.* p. 82.
11. J. Evelyn *A philosophical discourse of earth, at a meeting of the Royal Society, 24 June 1675* 1676, p. 26.
12. Giamatti *op cit.* p. 114.
13. *DWW* p. 318.
14. Williams *op. cit.* p. 74.
15. *PL* XII 464–5.
16. Quarles *op. cit.* Book IV no. 14, Book III no. 14.
17. Porro *op. cit.* 'Ai studiosi lettori'.
18. Bradley *Philosophical account* 1721, p. 181.
19. Parkinson *Paradisi in sole* p. 3.
20. Jacques Boyceau *Traité du jardinage* 1638, p. 71. Parkinson *Paradisi in sole* pp. 3, 5.
21. Evelyn's proposed Botanic Garden in EB pt. II ch. XVII. Temple *Works.* I p. 185. J. Worlidge *Systema horti-culturae or the art of gardening*, 1677, p. 18.
22. Austen *Spiritual use* proposition 94, pp. 452–68.
23. H. N. Clokie *An account of the herbaria of the department of botany in the University of Oxford* 1964, p. 1.
24. *The country gentleman and farmer's monthly director* 1726, p. ix; *general treatise* I p. 2.
25. de la Brosse *Description du jardin royal* p. 18. EB pt. II ch. XVII, p. 322.
26. *DWW* p. 288.
27. *PM* p. 287.
28. Stewart *op. cit.* p. 52.
29. *PM* p. 239.
30. EB pt. I ch. I.
31. Sir Thomas Browne *The Garden of Cyrus, or the quincuncial lozenge, or network plantations of the ancients, artificially, naturally, mystically considered* 1658, ch. I.
32. *EJB* p. 22.
33. Blunt *Linnaeus* p. 113. John Hill *An idea of a botanical garden in England* 1758. p. 13.
34. *PM* pp. 389–90.
35. John Ray *The wisdom of God manifested in the works of the creation* 2 parts 1692, pt. I pp. 9–10.
36. John A. Eddy 'The "Maunder minimum": sunspots and climate in the reign of Louis XIV' in Geoffrey Parker and Lesley M. Smith eds. *The general crisis of the seventeenth century* 1978.
37. Austen *Spiritual use* proposition 100, p. 493.
38. Raven *English naturalists* p. 248.
39. C. Webster 'The authorship and significance of Macaria' in C. Webster ed. *The intellectual revolution of the seventeenth century* 1974, p. 378.
40. Samuel Hartlib *A designe for plentie, by an universall planting of fruit trees* 1652, p. 4.
41. *Pomona* p. 2. Rohde *op. cit.* p. 158.
42. Timothy Nourse *Campania felix* 1700, p. 2.

43. Charles Evelyn *The lady's recreation* 1717, p. 138.
44. Rohde *op. cit.* p. 141.
45. Andrew Marvell *Upon Appleton house* stanza 78.
46. Thomas Whateley *Observations on modern gardening* 1801, p. 124: (first edn. 1770).
47. Moses Cook *The manner of raising, ordering, and improving forest trees* 1724, p. 183.
48. H. F. Clark *The English landscape garden* 1948, p. 6.
49. Hugh Prince *Parks in England* Shalfleet, Isle of Wight 1967, p. 4.
50. A. J. B. Le Blond *The theory and practice of gardening* tr. John James, 1712, p. 20. Bradley *Philosophical account* p. 181, *General treatise of husbandry and gardening* 2 vols 1726, II p. 248.
51. Stephen Switzer *Ichnographia rustica* 1718, p. xviii.
52. Batty Langley *New principles of gardening, or, the laying out and planting parterres, groves, wildernesses, labyrinths, avenues, parks &c., after a more Grand and Rural Manner than has been done before* 1728.
53. W. A. Brogden 'Stephen Switzer, "La grand manier" ' in *Furor hortensis, essays in memory of H. F. Clark* Edinburgh 1974, p. 23.
54. The Tradescants especially favoured large blooms, see Allan *op. cit.* p. 28. Evelyn *A philosophical discourse* p. 26.
55. *Isaiah* XL:4.
56. H. V. S. Ogden and M. S. Ogden *English taste in landscape in the seventeenth century* Ann Arbor 1955, p. 39.
57. G. Hakewill *An Apologie of the power and providence of God in the government of the world* &c, 1627, title page.
58. *PL* IV:256.
59. Ray *Wisdom of God*, pt. I p. 102, and pp. 199–206.
60. J. Beaumont *Considerations on . . . the theory of the earth* 1693, p. 87.
61. Bradley *General treatise* II, p. 242. James Thomson *The seasons* 1746 edn, Spring, lines 313, 320–2, 559–60; Summer, line 122.
62. Ogden and Ogden *op. cit.* pp. 39–40.
63. Beaumont *op. cit.* p. 87.
64. Ray *Wisdom of God* pt. I p. 11.
65. Le Blond *op. cit.* p. 20. Le Nôtre in C. Hussey *English gardens and landscapes 1700–1750* 1967, p. 22. Nourse *op. cit.* p. 4.
66. Ogden and Ogden *op. cit.* p. 40.
67. *ibid* p. 38, quoting *The shepheardes calender* (1579).
68. Joseph Beaumont *Psyche* (1648) quoted in J. G. Turner *op. cit.* p. 174.
69. Switzer *op. cit.* p. xv, Langley *op. cit.* title.
70. Temple *Works* I p. 186.

71. Prince *op. cit.* p. 6.
72. J. Laurence *Paradise re-gain'd, or the art of gardening* 1728, p. 16.
73. Thomas Beverley *Reflections upon the theory of the earth . . .*, 1699, pp. 14–15.
74. Hussey *op. cit.* p. 22.
75. Bradley *General treatise* II, p. 246.
76. Quoted in Hussey *op. cit.* pp. 28–9, and Røstvig *op. cit.* II p. 37.
77. For Pope see M. Mack *The garden and the city, retirement and politics in the later poetry of Pope 1731–1743* Toronto 1969, and Morris R. Brownell *Alexander Pope and the arts of Georgian England* Oxford 1978.
78. These lines, adapted from *Windsor Forest*, lines 15–16, appeared on the title page of B. Seeley's *Stow: a description of the gardens . . .*, 1744.
79. Clark *English landscape garden* pp. 6, 14.
80. Hussey *op. cit.* p. 29.
81. For Bridgeman see P. Willis *Charles Bridgeman and the English landscape garden* 1977.
82. *DWW* p. 198.
83. *PL* IV:341–7.
84. Horace Walpole *Essay on modern gardening* 1785 edn. pp. 27–33. For Brown see D. Stroud *Capability Brown* new edn. 1975.
85. Horace Walpole's epitaph in a letter to William Mason 10 February 1783, in E. Malins *English landscaping and literature 1660–1840*, 1966, p. 141.
86. Uvedale Price *An essay on the picturesque, as compared with the sublime and the beautiful &c.* new edn. 1796, p. 331.
87. Oliver Goldsmith *The deserted village* 1770.
88. Wilfrid Blunt *The Ark in the park, the zoo in the nineteenth century* 1976, p. 32.
89. *ibid* p. 25.
90. *ibid* p. 21.
91. *ibid* pp. 17–18.
92. *Dictionary of national biography*.
93. Alice Ford *Edward Hicks, painter of the peaceable kingdom* Philadelphia 1952, p. 86.
94. *ibid* p. 119.
95. Fritz Eichenberg *The wood and the graver: the work of Fritz Eichenberg* New York 1977, see especially, p. 25 *St Francis and the animals* (1934–6), p. 142 *The Peaceable Kingdom* (1950), and p. 13 *The Peaceable tree* (1977).
96. For a book covering some of these aspects together with some of those touched on here, see Terry Comito *The idea of the garden in the renaissance* Hassocks 1979.

BIBLIOGRAPHY

J. Q. Adams 'The wants of man', in William Harmon ed. *The Oxford book of American light verse* Oxford 1979.

Joseph Addison *The Spectator* nos 414, 25 June 1712, 477, 6 Sept. 1712. 'The pleasure of a garden', in A. H. Hyatt ed. *A book of old-world gardens* 1911.

Mea Allan *The Tradescants: their plants, gardens and museum 1570–1662* 1964.

D. C. Allen *The legend of Noah, renaissance rationalism in art, science and letters* Urbana 1949.

D. E. Allen *The naturalist in Britain, a social history* 1976.

K. Allott ed. *The poems of William Habington* 1948.

B. Ambrosinus *Panacea ex herbis quae a sanctis denominantur cocinnata opus* Bologna 1630.

Alicia Amherst (Hon. Mrs Evelyn Cecil) *A history of gardening in England* 1895, 3rd edn. 1910.

G. B. Andreini *L'Adamo, sacra rapresentatione* Milan 1617.

Anon. *The rise and progress of the present taste in planting parks, pleasure grounds, gardens &c, from Henry the eighth to king George the third* 1767.
——— *Anthologia, or the speach of flowers. Partly morall, partly misticall* 1655.

Agnes Arber *Herbals; their origin and evolution, a chapter in the history of botany, 1470–1670* Cambridge 1912.

H. Ashton *Du Bartas en Angleterre* Paris 1908.

G. Atkinson *La littérature géographique française de la Renaissance, répertoire bibliographique* Paris 1927.

Ralph Austen *A treatise of fruit-trees. Together with the spiritual use of an orchard* Oxford 1653.
——— *The spiritual use of an orchard or garden of fruit trees* second edn, Oxford 1657, repr. 1847.
——— *A dialogue, or familiar discourse and conference between the husbandman, and fruit trees, in his nurseries, orchards, and gardens* Oxford 1676.

P. Ildefonse Ayer 'Où plaça-t-on le paradis terrestre?' *Études Franciscaines* 1924.

F. Bacon *Works* ed. J. Spedding, R. L. Ellis, D. D. Heath, 14 vols 1857–74. *Of gardens* Ashendene Press 1897.

J. [T] Badeslade and J. Rocque *Vitruvius Britannicus* vol. IV 1739.

J. C. Barrell *The idea of landscape and the sense of place, 1730–1840* Cambridge 1972.

Joan Bassin 'The English landscape garden in the eighteenth century, the cultural importance of an English institution', in *Albion* 1979.

Mavis Batey 'Oliver Goldsmith, an indictment of landscape gardening' in P. Willis ed. *Furor hortensis* Edinburgh 1974.
——— *Nuneham Courtenay, Oxfordshire, a short history and description of the house, gardens and estate* Oxford 1970.

H. Baudet *Paradise on earth, some thoughts on European images of non-European man* New Haven 1965.

J. Bauhin *Historia plantarum universalis* 3 vols Ebroduni 1650–51.

J. Beale *Treatise on fruit trees* 1653.
——— *Herefordshire orchards* 1657.
——— *Nurseries, orchards, profitable gardens and vineyards encouraged . . .* 1677.

J. Beaumont *Considerations on a book, entituled the theory of the earth, publisht some years since by the learned Dr Burnet* 1693.

J. Beeverell *Les délices de la Grande Bretagne et de l'Irland* Leyden 1707.

A. T. Bellucio *Plantarum index horti Pisani* Florence 1662.

Julia S. Berrall *The garden. An illustrated history* Harmondsworth 1978.

T. Beverley *Reflections upon the theory of the earth, occasion'd by a late examination of it* 1699.

S. Blake *The compleat gardeners practice* 1664.

W. Blunt *The compleat naturalist, a life of Linnaeus* 1971.
——— *The ark in the park, the zoo in the nineteenth century* 1976.

G. Boas *Essays on primitivism and related ideas in the middle ages* Baltimore 1948.

H. Boerhaave *Index plantarum quae in horto academico Lugduno Batavo reperiuntur* Leyden 1710.
——— *Index alter plantarum quae in horto academico Lugduno – Batavo aluntur* Leyden 1720.

C. R. Boxer 'The Portuguese in the east 1500–1800' in H. V. Livermore ed. *Portugal and Brazil* Oxford 1953.
——— *Four centuries of Portuguese expansion, 1415–1825* Johannesburg 1961.
——— *The Dutch seaborne empire 1600–1800* 1965.
——— *The Portuguese seaborne empire, 1415–1825* 1969.

J. Boyceau *Traité du jardinage, selon les raisons de la nature et de l'art* Paris 1638.

R. Bradley *New improvements of planting and gardening* 1717, 1726 edn.
——— *A philosophical account of the works of nature endeavouring to set forth the several gradations remarkable in the mineral, vegetable and animal parts of the creation, tending to the composition of a scale of life* 1721, 1728 edn.
——— *A general treatise of husbandry and gardening* 2 vols 1726.
——— *The country gentleman and farmer's monthly director* 1726, 3rd edn 1727.

B. Braunrot *L'imagination poétique chez du Bartas, éléments de sensibilité baroque dans la création du monde* Chapel Hill 1973.

J. N. Brewer *The beauties of England and Wales* 1813.

W. A. Brogden 'Stephen Switzer, "La grand manier"', in P. Willis ed. *Furor hortensis* Edinburgh 1974.

A. Brome *Songs and other poems* 1661.

Wm. Brown, *see* P. Stephens.

Sir Thomas Browne *The garden of Cyrus, or the quincuncial lozenge, or network plantations of the ancients, artificially, naturally, mystically considered* 1658.

M. R. Brownell *Alexander Pope and the arts of Georgian England* Oxford 1978.

Sir E. A. W. Budge *The divine origin of the craft of the herbalist* 1928.

A. Bumaldus *Bibliotheca botanica seu herbariistarum scriptorum promota synodia* Bologna 1657.

J. Burmannus *Thesaurus Zeylanicus* Amsterdam 1737.

T. Burnet *Telluris theoria sacra* 1681.
———— *The theory of the earth* 2 vols 1684.

S. Cammau 'Types of symbols of Chinese art', in A. F. Wright ed. *Studies in Chinese thought* Chicago 1953.

J. M. F. Camp *Magic, myth and medicine* 1973.

Colen Campbell *Vitruvius Britannicus* vol. I 1715, II 1717, III 1725.

M. Carver *A discourse of the terrestrial paradise, aiming at a more probable discovery of the true situation of that happy place of our first parents habitation* 1666.

R. Castell *The villas of the ancients illustrated* 1728.

Hon. Mrs Evelyn Cecil, *see* Alicia Amherst.

Sir William Chambers *A dissertation on oriental gardening* 1772.

Isabel W. U. Chase *Horace Walpole: gardenist* Princeton 1943.

P. Chaunu *L'expansion européene du XIIIe au XVe siècle* Paris 1969.

C. M. Cippola *Guns and sails in the early phase of european expansion 1400–1700* 1965.

H. F. Clark 'Eighteenth century elysiums', in *England and the Mediterranean tradition* Oxford 1945.
———— *The English landscape garden* 1948.

Sir Kenneth Clark *Landscape into art* 1949.

George Clarke 'The Gardens of Stowe', in *Apollo* 1973.

D. Clifford *A history of garden design* 1962.

H. N. Clokie *An account of the herbaria in the department of botany in the University of Oxford* Oxford 1964.

D. R. Coffin *The Villa d'Este at Tivoli* Princeton 1960.

William Coles *The art of simpling, an introduction to the knowledge and gathering of plants* 1656.
———— *Adam in Eden: or, nature's paradise* 1657.

S. Collins *Paradise retriev'd: plainly and fully demonstrating the most beautiful, durable, and beneficial method of managing and improving fruit trees against walls, or in hedges, contrary to Mr Lawrence* 1717.

T. Comito *The idea of the garden in the renaissance* Hassocks 1979.

J. Commelin *The Belgick or Netherlandish hesperides, that is the management ordering, and use of the limon and orange trees* 1683.

M. Cook *The manner of raising, ordering, and improving forest-trees* 1676, 3rd edn 1724.

A. Cowley *Plantarum, the third and last volume of the works of Mr Abraham Cowley, including his six books of plants* 1721.

R. W. Crausius *Dissertatio inauguralis medica de signaturis vegetabilium* Jena 1697.

Sir Frank Crisp *Medieval gardens* 2 vols 1924.

G. R. Crone *Maps and their makers* 1953.
———— *The discovery of America* 1969.

A. W. Crosby *The Columbian exchange, biological and cultural consequences of 1492* Westport 1972.

N. Culpeper *The English physitian enlarged* 1656.

D'O. Dapper *Asia* Amsterdam 1672.
———— *Die unbekante neue Welt, oder Beschreibung des Welt-teils Amerika* Amsterdam 1673.
———— *Description de l'Afrique* Amsterdam 1686.

C. Daubeny *Oxford Botanic Garden: or, a popular guide to the botanic garden of Oxford* 2nd edn. Oxford 1853.
———— *A dream of the new museum* Oxford 1855.

I. David *Duodecim specula* Antwerp 1610.
———— *Paradisus sponsi et sponsae* Antwerp 1618.

Sir J. W. Dawson *Eden lost and won, studies of the early history and final destiny of man as taught in nature and revelation* 1895.

A. G. Debus *The English paracelsians* 1965.
———— *Science and education in the seventeenth century: the Webster–Ward debate* 1970.

Sergio Buarque de Holanda *Visão de paraíso* 2nd edn. Sao Paulo 1969.

G. de la Brosse *De la nature, vertu et utilité des plantes* Paris 1628.
———— *Description du Jardin Royal des plantes médicinales estably par le Roy Louis le juste à Paris* Paris 1636.

F. Delitzsch *Wo lag das paradies?* Leipzig 1881.

R. Dodoens *Histoire des plantes* tr. by C. de L'Escluse, Antwerp 1557.

R. Dodsley ed. *The works, in verse and prose, of William Shenstone* 3 vols, 5th edn 1777.

F. G. D. Drewitt *The romance of the Apothecaries Garden at Chelsea* 1922.

Saluste du Bartas *His divine weekes and workes* tr. by Joshuah Sylvester, 1605.
———— *Les oeuvres* Paris 1611.
———— *Part of du Bartas, English and French* tr. by William L'Isle, 1625.
———— *The works of Guillaume de Salluste, sieur du Bartas* ed. by U. T. Holmes, J. C. Lyons, R. W. Linker, 3 vols Chapel Hill 1935–40.

C. Duret *Histoire admirable des plantes et herbes esmerveillables et miraculeuses en nature* Paris 1605.

J. A. Eddy 'The "Maunder minimum": sunspots and climate in the reign of Louis XIV' in G. Parker and L. M. Smith eds *The general crisis of the seventeenth century* 1978.

F. Eichenberg *The wood and the graver: the work of Fritz Eichenberg* New York 1977.

M. Éliade 'The yearning for paradise in primitive tradition', in *Daedalus* 1959.

H. N. Ellacombe *The plant-lore and garden-craft of Shakespeare* 1896.

J. H. Elliott *The old world and the new, 1492–1650* Cambridge 1970.

R. J. W. Evans *Rudolf II and his world* Oxford 1973.

C. Evelyn *The lady's recreation, or the third and last part of the art of gardening improved* 1717.

J. Evelyn tr. *The French gardiner: instructing how to cultivate all sorts of fruit trees, and herbs for the garden: together with directions to dry and conserve them in their natural* 1658.

—— *Sylva* 1664.

—— *Pomona, or an appendix* [to *Sylva*] *concerning fruit trees in relation to cider* 1664.

—— *Kalendarium hortense* 1664.

—— *A philosophical discourse of earth, a paper read at a meeting of the Royal Society, 24 June 1675* 1676.

—— Elysium britannicum,' unpublished MS. in Christ Church Library, Oxford.

—— *Diary and correspondence of John Evelyn* ed. W. Bray, 4 vols 1850–2.

T. Fairchild *The city gardener, containing the most experienced method of cultivating and ordering such evergreens . . . exotick plants &c as will . . . thrive best in the London gardens* 1722.

B. Farrington *The philosophy of Francis Bacon* 1964.

J. Fletcher *The historie of the perfect-cursed-blessed man* 1628.

Alice Ford *Edward Hicks, painter of the peaceable kingdom* Philadelphia 1952.

J. Forster *England's happiness increased by a plantation of the roots called potatoes* 1664.

J. Frampton *Joyfull Newes out of the newe founde worlde, wherein is declared the rare and singuler vertues of diverse and sundrie hearbes, trees, oyles, plantes, and stones* taken from the Spanish of N. Monardes, 1577.

Rosemary Freeman *English emblem books* 1948.

Fruiterer's secret, The 1604.

J. Gerard *The herball or generall historie of plantes* 1597.

A. B. Giamatti *The earthly paradise and the renaissance epic* Princeton 1966.

William Gilpin *A dialogue upon the gardens of the Rt Hon. the Lord Viscount Cobham at Stow* 1748.

—— *Three essays on picturesque beauty, on picturesque travel, and on sketching landscape* 1792, 3rd edn 1808.

W. S. Gilpin *Practical hints upon landscape gardening: with some remarks on domestic architecture as connected with scenery* 1832.

M. L. Gothein *Geschichte der gartenkunst* 2 vols Jena 1914.

—— *A history of garden art* tr. from the German, 2 vols 1928.

D. B. Green *Gardener to Queen Anne: H. Wise, 1653–1738, and the formal garden* 1956.

T. M. Greene *The descent from heaven, a study in epic continuity* New Haven 1963.

R. T. Günther *Oxford gardens* Oxford 1912.

Wm. Habington. *see* Allott.

M. Hadfield *Gardening in Britain* 1960.

G. Hakewill *An apologie of the power and providence of God in the government of the world, or an examination and censure of the common errour touching nature's perpetuall and universall decay* 1627.

William Hanbury *An essay on planting, and a scheme for making it conducive to the glory of God, and the advantage of society* Oxford 1758.

—— *A complete body of planting and gardening* 2 vols 1770–71.

Sir Thomas Hanmer *The garden book of Sir Thomas Hanmer, bart.* with an introduction by E. S. Rohde, 1933.

H. Hare, Lord Coleraine, *The situation of paradise found out . . .* 1683.

T. Hariot *A briefe and true report of the newfoundland of Virginia* 1588.

J. Harris *Sir William Chambers* 1970.

S. Hartlib *A designe for plentie, by an universall planting of fruit trees* n.d.

—— *A discourse of husbandrie used in Brabant and Flanders* 1652.

H. Hawkins *Partheneia sacra* Rouen 1633.

H. Hediger *Wild animals in captivity* tr. from the German, 1950.

J. Heely *Letters on the beauties of Hagley, Envil and the Leasowes* 2 vols 1777.

P. Hermann *Paradisi Batavi prodromus sive plantarum exoticarum in Batavorum hortis observatarum index* Leyden 1689.

J. Hervey *Reflections on a flower garden* 1746.

P. Heylyn *Cosmographie* 1652.

Christopher Hill *Puritanism and Revolution* 1958.

John Hill *A general natural history . . .* 3 vols 1748–52.

—— *The useful family herbal* 1754.

—— *Thoughts concerning God and nature* 1755.

—— *Eden, or a compleat body of gardening* 1757.

—— *An idea of a botanical garden in England* 1758.

Oliver Hill and J. Cornforth *English country houses. Caroline 1625–1685* 1966.

R. B. Hinman *Abraham Cowley's world of order* Cambridge, Mass. 1960

C. C. L. Hirschfeld *Theorie der gartenkunst* Leipzig 1779–85.

W. G . Hiscock *John Evelyn and his family circle* 1955.

A. G. Hodgkiss *Discovering antique maps* 2nd edn. Aylesbury 1977.

H. Home, Lord Kames *Elements of criticism* 3 vols Edinburgh 1762.

P. H. Honour *The new golden land, European images of America from the discoveries to the present time* 1976.

William Howitt *The rural life of England* 2 vols 1838, 2nd edn 1840.

P. D. Huet *Traité de la situation du paradis terrestre* Paris 1691 (Eng. tr. 1694).

J. D. Hunt 'Emblem and expressionism in the eighteenth-century landscape garden', in *Eighteenth-Century Studies* III 1971.

—— and P. Willis *The genius of the place, the English landscape garden 1620–1820* 1975

—— *The figure in the landscape: poetry, painting and gardening during the eighteenth century* Baltimore 1976.

Husbandman's fruitful orchard, The 1608.

C. Hussey *English gardens and landscapes 1700–1750* 1967.

E. S. Hyams *The English garden* 1964.

—— *Great botanical gardens of the world* 1969.

—— *Plants in the service of man* 1971.

—— *Animals in the service of man* 1972.

A. H. Hyatt ed. *A book of old-world gardens* 1911.

B. D. Jackson 'A draft of a letter by John Gerard', in *Cambridge Antiquarian Communications* IV 1876–80, Cambridge 1881.

J. James *The theory and practice of gardening* tr. from the French of A. J. B. Le Blond, 1712 edn.

P. J. Jarvis 'The introduced trees and shrubs cultivated by the Tradescants at South Lambeth, 1629–1679', in *Journal of the Society for Bibliography of Natural History* 1979.

T. Johnson *Mercurius botanicus* 1634.

E. L. Jones 'The bird pests of British agriculture in recent centuries' in *Agricultural History Review* 1972.

J. Jonstone *An history of the wonderful things of nature* 1657 edn.

Margaret Jourdain *The work of William Kent* 1948.

Sir James Justice *The Scots gardiners director* Edinburgh 1754.

———— *The British gardener's new director, chiefly adapted to the climate of the northern countries* 4th edn Dublin 1765.

Mia C. Karsten *The old company's garden at the Cape and its superintendents* Cape Town 1951.

B. Kennicott *Two dissertations: the first on the tree of life in paradise* Oxford 1747.

J. Kip *Britannia illustrata* vol. I 1709, II 1716, III 1712.

R. P. Knight *The landscape, a didactic poem* 1794.

———— *The progress of civil society, a didactic poem* 1796.

P. Kolb *The present state of the Cape of Good Hope* 2nd English edn, 1738.

Margaret B. Kreig *Green medicine, the search for plants that heal* 1965.

Margaret W. Labarge *A baronial household of the thirteenth century* 1965.

B. Langley *New principles of gardening, or, the laying out and planting parterres, groves, wildernesses, labyrinths, avenues, parks &c., after a more Grand and Rural Manner than has been done before* 1728.

———— *Pomona: or, the Fruit-Garden illustrated* 1729.

E. Lankester *The uses of animals in relation to the industry of man* 1860–1.

J. Laurence *Paradise re-gain'd, or the art of gardening* 1728.

J. Lawrence *The clergyman's recreation* 1714.

———— *The gentleman's recreation* 1716.

William Lawson *A new orchard and garden* 1623, 6th edn 1683.

A. J. B. Le Blond, *see* J. James.

C. de L'Écluse *Rariorum plantarum historia* Antwerp 1601.

———— *Histoire des drogues espiceries, et de certains médicamens simples, qui naisent es Indes et en l'Amérique* Lyon 1619.

J. Lees-Milne *English country Houses. Baroque 1685–1715* 1970.

Ernst and Johanna Lehner *Folklore and odysseys of food and medicinal plants* 1973.

Leiden *Gids Hortus Botanicus Leiden* 1978.

H. Levin *The myth of the golden age in the renaissance* Bloomington 1969.

R. W. B. Lewis *The American Adam: innocence, tragedy and tradition in the nineteenth century* Chicago 1955.

C. Linnaeus *Musa Cliffortiana* Leyden 1736.

———— *Hortus Cliffortianus* Amsterdam 1737.

———— *Systema naturae* Stockholm 1740 edn.

F. Lippman *Zeichnungen von Sandro Botticelli zu Dante's Goettliche komoedie* Berlin 1884.

William L'Isle *Part of du Bartas, English and French* 1625.

H. V. Livermore ed. *Portugal and Brazil* 1953.

M. de Lobel *Plantarum seu stirpium historia* Antwerp 1576.

———— *Icones stirpium* Antwerp 1591.

G. Loisel *Histoire des ménageries de l'antiquité a nos jours* 3 vols Paris 1912.

J. C. Loudon *An encyclopedia of gardening* 1822.

A. O. Lovejoy *The great chain of being* Cambridge, Mass. 1936.

D. Lysons *The environs of London* 1792–6.

M. Macartney *English houses and gardens in the seventeenth and eighteenth centuries . . . reproduced from contemporary engravings by Kip, Badeslade, Harris and others* 1908.

M. Mack *The garden and the city, retirement and politics in the later poetry of Pope 1731–43* Toronto 1969.

Teresa McLean *Medieval English gardens* 1981.

P. Magnol *Botanicum Monspeliense* Montpellier 1686.

———— *Hortus regius Monspeliensis* Montpellier 1697.

E. Malins *English landscaping and literature 1660–1840* 1966.

Sir John Mandeville *The buke of John Maundevill* Roxburghe Club 1889.

K. Mannheim *Ideology and utopia* 1936.

E. W. Manwaring *Italian landscape in eighteenth century England* New York 1925.

G. Markham *The English husbandman* 1613, enlarged edn 1635.

———— *The country house-wife's garden* 1617, 1683 edn.

William Marshall *A review of 'the landscape', a didactic poem by R. P. Knight* 1795.

G. Mason *An essay on design in gardening* 1768.

William Mason *The English garden, a poem* 1772.

L. G. Matthews 'Herbals and formularies' in F. N. L. Poynter ed. *The evolution of pharmacy in Britain* 1965.

P. A. Mattioli *Commentarii, in libros sex pedacii Dioscoridis . . . de medica materia* Venice 1554.

L. Meager *The English gardner* 1670, 1699 edn.

O. F. Mentzel *A geographical and topographical description of the Cape of Good Hope* tr. by H. J. Mandelbrote, 2 parts Cape Town 1921, 1925.

K. Miller *Mappaemundi. die ältesten weltkarten* Stuttgart 1895.

P. Miller *The gardeners and florists dictionary, or a complete system of horticulture* 2 vols 1724.

J. Mitford *The correspondence of T. Gray and William Mason* 1853.

S. E. Morison *Admiral of the ocean sea, a life of Christopher Columbus* 2 vols Boston 1942.

W. O. Newnham *Alresford essays for the times* 1891.

Marjorie H. Nicolson *Newton demands the muse:*

Newton's opticks and the eighteenth century poets Princeton 1946.

T. Nourse *Campania felix* 1700.

H. V. S. and M. S. Ogden *English taste in landscape in the seventeenth century* Ann Arbor 1955.

M.A.E.P.P. *Catalogus plantarum singularum suis areolis distinctarum scholae botanicae horti regii Parisiensis...*, Paris 1656.

P. Paaw *Hortus publicus Academiae Lugduno-Batavae...* Leyden 1601.

Padua *Guide to the botanical gardens founded in 1545* Padua 1973.

E. Panofsky *Meaning in the visual arts* New York 1955.

Paris *Plantarum index* Paris 1661.

T. Park ed. *The poetical works of Nathaniel Cotton* 1806.

J. Parkinson *Paradisi in sole, paradisus terrestris* 1629.
——— *Theatrum botanicum* 1640.

J. H. Parry *The age of reconnaissance 1450–1650* 1963.
——— *The Spanish seaborne empire* 1966.

H. R. Patch *The other world, according to descriptions in medieval literature* Cambridge, Mass. 1950.

R. Paulson *Emblem and expression, meaning in English art of the eighteenth century* 1975.

P. Pawi, *see* Paaw.

G. Pellissier *La vie et les oeuvres de du Bartas* Paris 1882.

B. Penrose *Travel and discovery in the renaissance, 1420–1620* Cambridge, Mass. 1952.

J. L. Phelan *The millenial kingdom of the Franciscans in the new world, a study of the writings of Geronimo de Mendieta (1525–1604)* Berkeley and Los Angeles 1956.

T. G. Pinches *The Old Testament, in the light of the historical records and legends of Assyria and Babylonia* 1902.

Sir Hugh Plat *The garden of Eden, or, an accurate description of all flowers and fruits now growing in England* 1653.

R. Poggioli 'The pastoral of the self', in *Daedalus 1959*

J. Pomet *Le marchand sincère ou traité général des drogues* Paris 1695.

G. Porro *L'horto de i semplici di Padova* Venice 1591.

F. N. L. Poynter ed. *The evolution of pharmacy in Britain* 1965.

E. Prestage *The Portuguese pioneers* 1933.

U. Price *An essay on the picturesque, as compared with the sublime and the beautiful &c.* new edn 1796.

H. Prince *Parks in England* Shalfleet, Isle of Wight 1967.

R. E. Prothero 'Agriculture and gardening' in *Shakespeare's England* Oxford 1916.

F. Quarles *Emblemes* 1635.

Sir Walter Raleigh *The history of the world* 1614.

J. P. Rameau *Démonstration du principe de l'harmonie* Paris 1750.

C. E. Raven *John Ray, naturalist* Cambridge 1942, 1950 edn.
——— *English naturalists from Neckham to Ray* Cambridge 1947.

Ray Society, The *Memorials of John Ray* ed. E. Lankester 1846.

J. Ray *Catalogus plantarum Angliae* 1670.
——— *Observations . . . made in a journey through part of the Low-Countries, Germany, Italy and France; with a catalogue of plants not native of England, found spontaneously growing in those parts, and their virtues* 1673.
——— *The wisdom of God manifested in the works of the creation* 1692 edn.

J. Rea *Flora: seu de florum cultura* 1665.

K. Reichenberger *Die schöpfungswoche des du Bartas. kritischer text* Tübingen 1963.

H. Repton *Observations on the theory and practice of landscape gardening* 1803.
———*An inquiry into the changes of taste in landscape gardening* 1806.
——— *Fragments . . .* 1816.

F. E. Robbins *The hexaemeral literature, a study of the Greek and Latin commentaries on Genesis* Chicago 1912.

P. A. Robin *Animal lore in English literature* 1932.

M. S. Røstvig *The happy man* 2 vols, 2nd edn. Oslo 1962.

E. S. Rohde *The old English gardening books* 1924.
——— *Garden craft in the Bible and other essays* 1927.
——— *The story of the garden* 1932.

L. Rumetius *Scripturae sacrae viridarium literale et mysticum* Paris 1626.

C. L. Sanford *The quest for paradise, Europe and the American moral imagination* Urbana 1961.

A. H. Sayce *The 'higher criticism' and the verdict of the monuments* 1894.
——— 'The trees of life and knowledge', in *Florilegium* ed. Melchior de Vogue, Paris 1909.

J. W. Schiffeler 'Chinese rock gardens: the roots of heaven', vol. VIII *Journals of the China Society* 1971 Taipei.

G. Schomberg *British zoos, a study of animals in captivity* 1957.

H. Schultz *Milton and forbidden knowledge* New York 1955.

F. Schuyl *Catalogus plantarum horti academici Lugduno Batavi* Leyden 1668.

B. Seeley *Stow: a description of the gardens of the rt. hon. the Lord Viscount Cobham* Northampton 1744.

R. B. Serjeant *The Portuguese off the south Arabian coast* Oxford 1963.

J. Serle *A plan of Mr Pope's garden as it was left at his death . . .* 1745.

William Shenstone, *see* R. Dodsley.

J. C. Shepherd and G. A. Jellicoe *Italian gardens of the renaissance* 1925, 1966 edn.

C. S. Singleton *Dante studies* Cambridge, Mass. 1 1954, 2 1958.

O. Siren *China and gardens of Europe* New York 1950.

R. A. Skelton *Explorers maps* 1958.

B. Snell *The discovery of the mind: the Greek origins of European thought* tr. Oxford 1953.

A. Speed *Adam out of Eden, or, an abstract of divers excellent experiments touching the advancement of husbandry* 1659.

L. Spitzer 'Classical and christian ideas of world harmony' in *Traditio* 1944, 1945.

C. Stengel *Hortensius, et Dea Flora, cum Pomona historice, tropologice, et anagogice descripti* Augsburg 1647.

P. Stephens and W. Brown *Catalogus Horti Botanici Oxoniensis* Oxford 1658.

Sir Henry Steuart *The planters guide* Edinburgh 1828.

S. Stewart *The enclosed garden, the tradition and the image in seventeenth-century poetry* Madison 1965.

P. Street *The London zoo* 1956.

R. Strong *The cult of Elizabeth, Elizabethan portraiture and pageantry* 1977.
———— *The renaissance garden in England* 1979.

Dorothy Stroud *Humphrey Repton* 1962.
———— *Capability Brown* new edn. 1975.

H. E. Stutchbury *The architecture of Colen Campbell* Manchester 1967.

J. Swan *Speculum mundi, or, a glasse representing the face of the world* Cambridge 1635, 3rd edn. 1665.

S. Switzer *Ichnographia rustica. The nobleman, gentleman, and gardener's recreation* 1718.

J. Sylvester tr. du Bartas *His divine weekes and workes* 1605.

G. Tachard *Voyage de Siam des pères Jésuites envoyez par le roy aux Indes et à la Chine* Paris 1686

G. C. Taylor *Milton's use of du Bartas* Cambridge, Mass. 1934.

Sir William Temple *Upon the gardens of Epicurus, or of gardening in the year 1685* in *Works* 2 vols 1720 edn.

C. Thacker 'Voltaire and Rousseau: eighteenth century gardeners', in *Studies on Voltaire and the eighteenth century* XC 1972.

K. V. Thomas *Religion and the decline of magic . . .* 1971.

F. W. Thompson *A history of Chatsworth* 1949.

'Tradescant's orchard', Ashmole Ms. 1461 Bodleian Library, Oxford.

J. Trusler *Practical husbandry; or, the art of farming, with a certainty of gain* 1780.
———— *Elements of modern gardening; or, the art of laying out pleasure grounds* 1784.

T. Tryon *A treatise of cleanness in meats and drinks . . .* 1682.
———— *The way to health, long life and happiness &c.* 1683.
———— *The way to make all people rich, or wisdom's call to temperance and frugality* 1685.
———— *The good housewife made a doctor . . . .* c. 1690.

J. G. Turner 'Topographia and the topographical poem in English 1640–1660', D. Phil. thesis, Oxford University 1976.
———— *The politics of landscape, rural scenery and society in English poetry 1630–1660*, Oxford 1979.

P. J. Ucko and G. W. Dimbleby eds *The domestication and exploitation of plants and animals . . .* 1969.

P. Vallet *Le jardin du roy . . . Henri IV* Paris 1608.

F. van Sterbeeck *Citricultura* Antwerp 1712.

S. van Til *Dissertationes philologico-theologicae* Leyden 1719, 1744 edn.

H. Veendorp and L. G. M. Baas Becking *1587–1937 Hortus Academicus Lugduno Batavus: the development of the Gardens of Leyden University* 1938.

Versailles *La description du chateau de Versailles* 1694.

Vertumnus *An epistle to Mr Jacob Bobart* Oxford 1713.

J. Veslingio *Catalogus plantarum horti gymnasi Patavini* Padua 1644.

H. Walpole *Essay on modern gardening* Strawberry Hill 1785.

E. Warren *Geologia: or, a discourse concerning the earth before the deluge, wherein the form and properties ascribed to it in a book intituled 'The theory of the earth', are excepted against* 1690.

William Watts *The seats of the nobility and gentry* 1779.

C. Webster 'The authorship and significance of Macaria' in C. Webster ed. *The intellectual revolution of the seventeenth century* 1974.
———— *The great instauration, science, medicine and reform, 1626–1660* 1975.

M. Welch *The Tradescants and the foundation of the Ashmolean Museum* Oxford 1978.

T. Whateley *Observations on modern gardening* 1770, 1801 edn.

L. Whistler *The imagination of Vanbrugh and his fellow artists* 1954.

J. J. Wilhelm *The cruelest month, spring, nature, and love in classical and medieval lyrics* New Haven 1965.

L. P. Wilkinson *Horace and his lyric poetry* Cambridge 1945.

Arnold Williams *The common expositor. An account of the commentaries on Genesis 1527–1633* Chapel Hill 1948.

G. H. Williams *Wilderness and paradise in christian thought* New York 1962.

P. Willis 'Rousseau, Stowe and le jardin anglais: speculations on visual sources for la nouvelle Héloise', in *Studies on Voltaire and the eighteenth century* XC 1972.
———— ed. *Furor Hortensis, essays in memory of H. F. Clark* Edinburgh 1974.
———— *Charles Bridgeman and the English landscape garden* 1977.

G. Wither *A collection of emblemes* 1635.

A. Wolf *A history of science, technology and philosophy in the sixteenth and seventeenth centuries* 1950.

K. Woodbridge 'Henry Hoare's paradise', in *The Art Bulletin* 1965.
———— *Landscape and antiquity, aspects of English culture at Stourhead 1718–1838* Oxford 1970.

J. Worlidge *Systema horti-culturae or the art of gardening* 1677.
———— *A compleat system of husbandry and gardening, or the gentleman's companion in the business and pleasures of a country life* 1716.

A. F. Wright *Studies in Chinese thought* Chicago 1953.

Gilette Ziegler *The court of Versailles in the reign of Louis XIV* tr. 1966.

INDEX